GYMPIE GOLD

HECTOR HOLTHOUSE

Angus&Robertson
An imprint of HarperCollins*Publishers*, Australia

First published in Australia by Angus&Robertson Publishers in 1973
Queensland Classic edition 1980
This Angus&Robertson Classics edition published in 1999
by HarperCollins*Publishers* Pty Limited
ACN 009 913 517
A member of the HarperCollins*Publishers* (Australia) Pty Limited Group
http://www.harpercollins.com.au

Copyright © The Estate of Hector Holthouse 1973

This book is copyright.
Apart from any fair dealing for the purposes of private study, research, criticism or review, as permitted under the Copyright Act, no part may be reproduced by any process without written permission.
Inquiries should be addressed to the publishers.

HarperCollins*Publishers*
25 Ryde Road, Pymble, Sydney, NSW 2073, Australia
31 View Road, Glenfield, Auckland 10, New Zealand
77–85 Fulham Palace Road, London, W6 8JB, United Kingdom
Hazelton Lanes, 55 Avenue Road, Suite 2900, Toronto, Ontario M5R 3L2
and 1995 Markham Road, Scarborough, Ontario M1B 5M8, Canada
10 East 53rd Street, New York NY 10032, USA

National Library of Australia Cataloguing-in-Publication data:

Holthouse, Hector.
  Gympie gold.
  Includes bibliography.
  Includes index.
  ISBN 0 2071 9693 1
  1. Gold mines and mining – Queensland – Gympie – History.
  2. Gympie, (Qld.) – History.
  I.Title
994.32

Cover photograph: Stock Photos Photographic Library
Printed in Australia by Griffin Press Pty Ltd, on 79gsm Bulky Paperback

5 4 3 2 1
03 02 01 00 99

Foreword

"Heritage of the Past – Mining for the Future"

This book was reprinted in 1999 with the assistance of Gympie Gold Limited, a public, listed gold mining company whose wholly-owned subsidiary, Gympie Eldorado Gold Mines Pty. Ltd., is referred to on the last page of this book as reporting gold mineralisation in the El Dorado shaft at Gympie's Albert Park sports oval in 1971.

This mineralisation proved uneconomic at that time but it encouraged Gympie Eldorado Gold Mines to amalgamate the Gympie Goldfield leases in the early 1970s. Exploration drilling commenced in 1980 and discovered zones of high grade ore at depth. In 1988, the deepest shaft, West of Scotland in southern Gympie, was reopened after 84 years of dormancy. The field was de-watered and the Scottish Gympie No. 2 Shaft was reopened in 1996. Ventilation was upgraded by an exhaust fan fitted to the Scottish Gympie No. 3 Shaft. In June 1988, the old Scottish Gympie No. 1 Shaft was reopened and renamed the Bas Lewis shaft and in September 1998 refurbishing of the No. 2 South Great Eastern Shaft at the Mining Museum commenced. Over $70 million has been invested to establish the modern Monkland Mine which operates from 400 to 900 metres depth (1,300 to 3,000 feet).

The Monkland Mine produces over 30,000 ounces of gold per annum from its gold mill located at Widgee Gully. It also owns tenements covering the Gympie Goldfield and surrounding exploration areas exceeding 1,300 square kilometres (500 square miles). The company intends to continue expanding its operations through exploration and development, reclaiming its former standing as as a world-class goldfield with several mines supporting each other in a co-ordinated organisation.

The Gympie gold mine operates successfully beneath a modern city with the support of the local community. Gympie developed over the extensive historical mine workings and during the 1990s the Queensland Department of Mines and Energy began capping the old shafts. Apart from old shafts, deep gold mines rarely impact on surface conditions due to the relatively small openings and hard rock.

In 1997, the Monkland Mine's first modern "jeweller's shop" ore was found and spectacular gold specimens began to be marketed internationally, bringing recognition to Gympie.

The 1998 Gympie Eldorado workforce numbered 110 persons. And just as in the historical era, for every miner, there are several others in the town benefiting from the business activity, employment and optimism that has been a hallmark of all Gympie gold mines since 1867. Long may this continue.

GYMPIE GOLD LIMITED

Modern Headframe
at Scottish Gympie No. 2 Shaft.
Part of the Monkland Mine.

Contents

1	Stinging-tree Country	1
2	Canoona's Black Nuggets	9
3	The Battle of Crocodile Creek	18
4	The Loner from Calliope	25
5	Sitting on Gold	31
6	First on the Field	39
7	Mother of Gold	52
8	The Escort Murder	59
9	Shanty Town Christmas	69
10	The Big Nugget	78
11	Women in the Wet	86
12	The Carnival Year	96
13	Chinaman's Flat	105
14	Breaking Out the Quartz	111
15	Bail Up!	126
16	Bushranger's Conscience	140
17	Bringing in the Big Cake	154
18	Black Rebel	167
19	"Jeweller's Shop" Days	178
20	The Great Flood	186
21	Golden Dividends	193
22	Green Pastures	198
Bibliography		201
Index		203

Acknowledgments

The author wishes to thank the staff of the Oxley Memorial Library, Brisbane, for their help in obtaining material used in this book.

Photographs by Richard Daintree, geologist, squatter and later Queensland Agent-General in London, are from originals held by the Oxley Memorial Library.

Author's Note

In compiling this history, which is based partly on the reports and reminiscences of Gold Commissioners, geologists, miners and their families, I have sometimes edited quotations to remove redundant material but have been careful not to change the meaning. Dialogue used in the book is either quoted direct from such sources or is based on descriptive material.

List of Illustrations

James Davis (Duramboi)
James Nash, who discovered the Gympie Goldfield
Digging for alluvial gold in shallow gravel
Cradling for alluvial in a scrub-surrounded gully
Rough bough shelters cover the shafts
Tents and rough bark shanties beside a gold-bearing gully
Mullock heaps among the tall timbers
Gympie at the height of the rush in the 1870s
Mary Street, Gympie, at the end of 1868
Mary Street, still unpaved, in the early 1870s
Cartoonist's impression of diggers' dry humour
Horse-powered whip
Whim used for raising ore from the mines
Gympie under water during the flood of 1893
William Couldrey
Matthew Laird
Remains of the Scottish Gympie mine
Victoria House, home of Jacob Pearen
Gympie's last big mullock heap
Old stampers, now part of the Gympie Museum
Mary Street, Gympie, today
Gympie's Town Hall, with the monument to James Nash
James Nash's grave in Gympie Cemetery
Gympie's Memorial Park, with a monument to James Nash
Gympie's new El Dorado mine
Phoenix Reborn mine

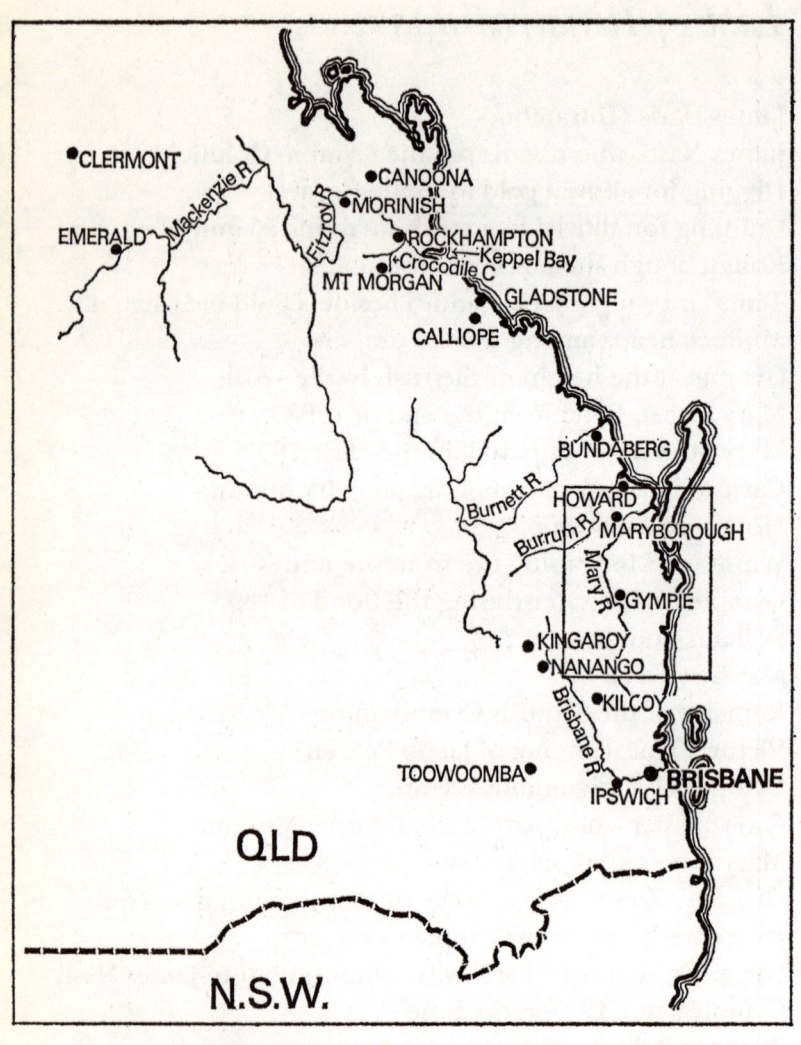

SOUTH-EAST QUEENSLAND

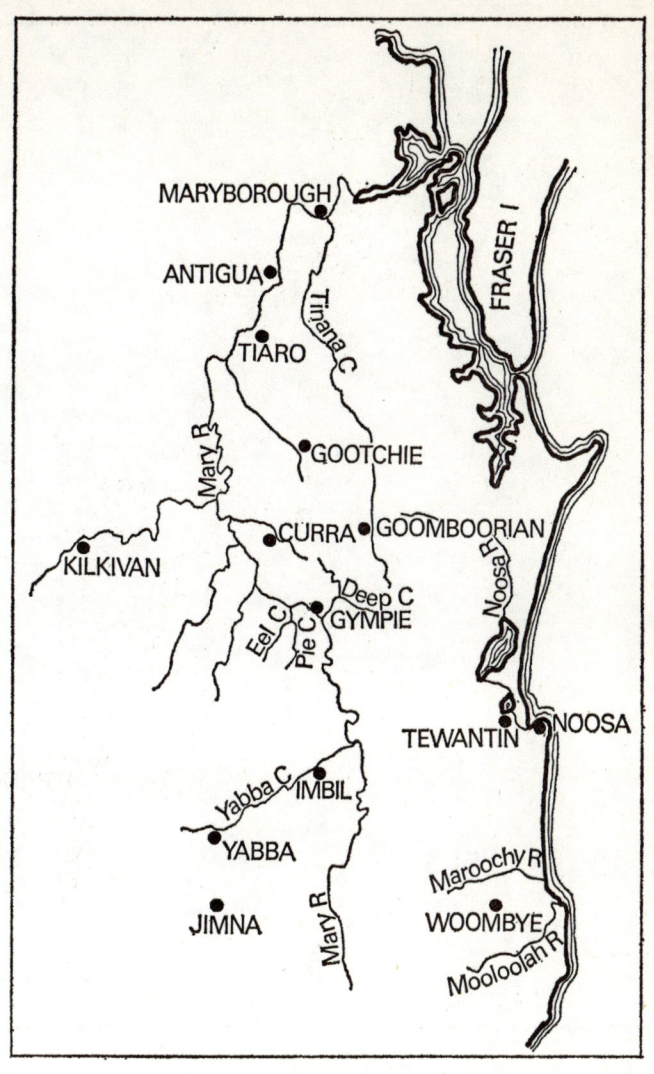

GYMPIE AND SURROUNDING DISTRICTS

Chapter One:
Stinging-tree Country

The lone prospector, weary from a long day's trudge, sat slumped against an outcrop of brown rock streaked with bands of white quartz. There were others like it all around him, and in the last of the daylight he could see a line of white running along the ridge above him, like a backbone, and patches dotting the slopes right down to the gully where he had been digging. His dog lay about three feet away, fed and ready to sleep, but watching him to see what he would do. The animal sensed his master's uneasiness.

Slowly and methodically the man's eyes swept the scrubby slopes of surrounding ridges and tried to penetrate the darkening shadows of the thick, tangled jungle that lined the banks of the slow-flowing river a few hundred yards away. It was hostile jungle, full of the stinging Gympie-Gympie tree that could blister the skin off a man at a touch, but for the first time in his life he was grateful for the Gympie-Gympie. The thicker the stinging-tree grew the better he liked it now.

Once again the prospector looked down at the heavy pebbles in the palm of his hand. They were gold. There was gold all around him—gold in the ridges, gold in the quartz he leaned against, gold in the gravel of the gully.

He picked up his blackened billy, drained the last of the almost cold tea it contained, and paused in the act of throwing away the leaves. Then he knocked them out carefully over the dying embers of his fire and picked up a stick and scratched them around in the dirt until he was sure the fire was out. With darkness coming, he wanted no tell-tale glow to show some other prospector or stockman there was somebody here.

That was what gold did to a man. Last night any other human being he met on the track would have been a mate to stop and yarn with. But that was before he found the gold. When a man found gold the first thing he did was look over his shoulder to make sure nobody had seen him. Jimmy Nash wanted no mates now. He had searched long enough for this gold he had found in the stinging-tree country. He wanted nobody else coming along to share it just yet.

Nash put his gold into a small leather bag he had carried all his prospecting life and shoved it into the pocket of his moleskin trousers. He made sure the fire was quite out, spread his blanket on a level patch of ground, and lay down and wrapped it about him. The dog dropped his head on his paws and relaxed.

It was late Sepember 1867 when prospector James Nash lay down with his gold near the banks of the Mary River, about a hundred miles north of Brisbane. What he had found was going to begin the eight-year-old colony of Queensland's first big gold rush, and the banks of the Mary River, as yet unseen by any but a few cedar-getters and stockmen, were soon to be stripped to the bones by hordes of hungry gold-seekers.

The Mary rises in ranges less than fifty miles north of Brisbane and winds north for nearly two hundred miles before swinging east into the sea behind the cover of the great sandbank called Fraser Island. It meanders through hilly country once rich in huge red cedar trees, tall pines straight as gun barrels, gums and ironbarks. In the ranges it is fringed with jungle; its lower reaches are bordered by fertile, alluvial flats. Before the coming of the white man its whole valley was good country swarming with game and the home of virile and warlike Aborigines.

The tribes have been given various names by different observers who found difficulty in distinguishing between them, but all were probably related to the Kabis, a well-built, vigorous people whose territory extended north from Noosa and a considerable distance inland.

The first white men to penetrate this country where James Nash was later to find his El Dorado were runaway convicts who lived with the Aborigines for considerable periods before returning or being brought back to Brisbane. The one to get the farthest afield was James Davis, probably aged about twenty-four (though some authorities make it several years younger) when he escaped from the Moreton Bay settlement in March 1829 and headed north. Treated tolerantly by the Kabis, he made his way to the headwaters of the Mary and followed it down to the territory of a tribe generally referred to as the Ginginburras, where one of the old men claimed him as his dead son, Duramboi, come back to life, and adopted him.

By 1841 settlers were pushing north along the headwaters of the Brisbane River to establish Cressbrook Station around present-day Toogoolawah, Kilcoy Station in the Stanley River valley, and Durundur Station to the east of it. Others were out looking for land.

On 4 May 1842 Andrew Petrie, Foreman of Works at Moreton Bay and spare-time explorer, together with landseekers Henry Stuart Russell, W. Wrottesley and a former Royal Navy midshipman named Joliffe, left Moreton Bay in a gig manned by a crew of convict rowers to explore the coastal country north of Brisbane. They discovered and entered the Mary River, found Duramboi living with the Ginginburras somewhere near the present township of Tiaro, and induced him to come back to Brisbane with them.

Joliffe gave a good report of the country to his employer, John Eales of the Hunter River district, and was sent overland from Moreton Bay with 20,000 sheep, several drays, and Duramboi as guide to establish a run for Eales on the newly discovered river near Tiaro. The schooner *Edward* was sent up the river to bring in supplies.

Nobody had taken much notice of a warning by Duramboi that, following the poisoning of about sixty Aborigines on Kilcoy Station, the Kabis were seething with resentment and organizing warfare against all settlers on a wide scale. Joliffe soon found the warning was well based. Two of his shepherds were speared, sheep were driven off and slaughtered, and soon no white man dare venture from cover without running the risk of being speared from ambush. Sheep scattered while shepherds sheltered in their huts, prevented from deserting only by the fact that they were surrounded by the enemy.

Joliffe resigned and was replaced by a Mr Last who fared no better. More men were speared and eventually the run had to be abandoned, leaving the sheep to the Ginginburras who invited their neighbours from hundreds of miles around to a feast and corroboree that was to be remembered for years.

Duramboi told Russell there was better land to the north. Russell went looking for it and took up a promising run on the Boyne River, a tributary of the Burnett which drained the country north of the Mary valley. Others who followed also skirted the dangerous Mary River country, but inevitably they gradually closed in on it.

Early in 1847 timber-getter George Furber arrived from Ipswich, took up part of Eales's abandoned run and built a shanty and wharf on the southern bank of the lower reaches of the Mary where Tinana Creek joined it. He advertised in the *Moreton Bay Courier* in June 1847 that he was opening a store for the reception of wool at the head of navigation of the "Wide Bay River", the name of both river and district being taken from Wide Bay at the south-eastern end of Fraser Island.

In July 1847 surveyor J. C. Burnett surveyed the Wide Bay River and, on Governor FitzRoy's instructions, named it the Mary after the Governor's wife, Lady Mary FitzRoy. He named the next big river to the north the Burnett after himself, and further north again, named the Fitzroy River —also after the Governor's wife. Prospective settlers scouting the country at the time were enthusiastic. Settler-historian George Loyau wrote:

> The whole Wide Bay district was then an uninterrupted meadow of waving grass on which the few stock could scarcely make an impression. Kangaroos and emus hopped or ran lazily across the plains; ducks flew out of every waterhole and black swan was often met with; whilst the stately bustard strutted and puffed out its

breast feathers as if indignant at the intrusion. Every tree appeared full of foliage and the locust sang gleefully as if to welcome the white man to the New Arcadia.

If the locusts felt that way about it, they must have been the only living things that did. The Aborigines certainly did not.

The Ginginburras and their neighbours had not changed their opinion of white men; in October 1847, as Furber and an employee named Barron were building a fence, two Aborigines they were employing sneaked up behind them and as one split Furber's skull with a squaring axe the other brained Barron with a mortising axe. Furber survived and on regaining consciousness, in spite of his terrible wound, managed to catch a horse and ride 150 miles to Ipswich. He recovered, returned, and shot the man who had hit him. But the Ginginburras had long memories, and they ended the feud in 1855 by killing both Furber and his son-in-law William Wilmhurst.

In the meantime, however, Furber's wharf at Tinana became a rallying point for shearing and the boiling down of flocks from distant stations, communications with the south being kept up by a small trading schooner, the *Aurora*. Eventually the settlement, by then known as Maryborough, was moved to the present site about twenty-five miles up-river from the sea.

Settlers were now pushing into the valley determined to take the fertile land, and there was nothing that was going to stop them. The warfare that followed was bitter and ruthless. "Every acre of land in these districts was won from the Aborigines by bloodshed and warfare, whilst in some instances poison played an important part," wrote Loyau. "No white man was safe without his rifle or Colt revolver at hand. The jungle on the banks of the Mary enabled them to move through the country without being seen, and now and then the settler discovered their presence in his vicinity by a dead bullock or two with spears in the carcass or in a report from the out station of some unfortunate shepherd or hut keeper having got a bad headache with a nulla-nulla or tomahawk."

Homesteads were built like miniature fortresses. The

original Imbil homestead, near Yabba Creek, an upper tributary of the Mary, was described by William Everett who pulled it down.

> It had been built when there was no sawn timber to be had. It consisted wholly of round timber, the top and bottom plates being large saplings, adzed on the top and underneath. Then there was a two-inch groove cut the full length. A piece was cut out at each end and the slabs were slid along. The floor joist was saplings faced on the top side and all corner posts were round timber. The slabs were all cedar, being twenty-seven inches in width. There were a lot of big auger holes through the slabs for the people to shoot through if the blacks made a raid.

The cedar-getters had moved into the valley with the first settlers, and the trees, as well as being used for most local building purposes, were dragged to the river by bullock teams and floated downstream for recovery and shipment south to Melbourne. Some of these men found traces of gold and talked about it, but the gold with which they were concerned came from the centuries-old trees they were cutting out so ruthlessly. None of them stopped to scratch in the ground, and nobody else was interested enough in their yarns to go and see for himself.

Even men who should have known better ignored the gold that lay waiting in the Gympie-Gympie scrubs.

In 1848 John Carne Bidwill, a distinguished British botanist, left at a loose end by a Government misunderstanding in Sydney, was appointed first Commissioner of Crown Lands for the Wide Bay area. He built a home on Tinana Creek and planted what he vainly hoped would become a botanic garden, but he had little chance to enjoy either.

The New South Wales Government, in order to give Maryborough an overland route to Brisbane, instructed Bidwill in 1852 to survey a marked tree line from Wide Bay to Durundur Station. A track from Durundur to Brisbane had already been opened.

Bidwill's line south from Maryborough went through thickly timbered country with alternate belts of hardwood ridges and valleys of dense scrub watered by the Mary River and dozens of creeks and gullies that ran into it. Much of this was stinging-tree country. Gympie-Gympie was an Aboriginal expression supposed to mean something like "guardian of evil spirits" and was applied to a tree, later classified as the most virulent member of the *Laportea* genus of stinging-trees, of which Australia has three species. It was more a shrub than a tree, rarely exceeding ten feet in height, and its stem and the underside of its large, oval leaves were covered with stiff, stinging hairs carrying a powerful poison containing formic and acetic acid. A bad sting could cause days of excruciating pain, and the after-effects lasted for months.

A large part of Bidwill's trip was very heavy going, with many running creeks to be forded, tracks to be cut through Gympie-Gympie scrub, and a constant watch kept against possible attack by Aborigines. At one spot in a natural clearing in the scrub the party came on a huge heap of wool from at least two thousand sheep. The Aborigines had run the stolen sheep through the scrub and held them there for weeks of feasting. The meat had been eaten and the skins scraped clean, but for the wool the Aborigines had no use, and there it remained.

Accompanying Bidwill on the trip was pioneer George W. Dart who had landed at Tinana wharf in August 1850. Long after the goldrush that Nash started had come and gone and the town of Gympie had grown up on the field, Dart wrote:

> The first gold discovered in Gympie was found by Mr J. C. Bidwill. He followed up a spur of the main range that divided the Tinana waters from the Mary River watershed, and he continued his marked tree line right into Gympie; and it was while he was delayed at Gympie making a temporary bridge over a creek there that he found gold. Bidwill's marked tree line went straight through Gympie. I saw the gold myself. Mr Bidwill was in the habit of showing the gold to many of his friends from the country when they stayed there for dinner.

Soon after the party left the place where the gold was found rations ran short and Bidwill and a man named Slade went on ahead to try to reach Durundur. They got lost, their food ran out, and they lived on roots. For eight days they wandered around the bush, not knowing where they were, and getting weaker daily. They were almost dead when found by a group of station Aborigines and taken to Durundur. The tree line was finished by surveyor Buchanan and in consequence was generally called Buchanan's line.

Bidwill never recovered from the privations of the trip, and he died at Tinana on 16 March 1853, aged thirty-eight. He is remembered today by the scientific name of the stately Bunya Pine (*Araucaria bidwillii*), but not as the first white man to find the stinging-tree gold.

Chapter Two:
Canoona's Black Nuggets

In the fifteen years that followed Bidwill's discovery of gold, many found traces of gold in the stinging-tree country but none made a thorough search, and when the first big gold strike was made in the north it was not on the Mary River, but far beyond it on the Fitzroy.

The Archer brothers, founders of Durundur Station, had found that area unsuitable for sheep, and moved north to take up Gracemere Station, on the north-western side of the Fitzroy River, in 1855. They built a woolshed on the river bank and ran their schooner, the *Jenny Lind*, to take out their wool and bring in supplies. The following year the Elliotts took up Canoona Station, to the north of the river, and Richard Palmer set up a general store beside the Fitzroy and built a wharf. By 1857 there were enough thirsty station hands and teamsters in the district for R. A. Parker to build a slab and bark shanty which he called the Bush Inn.

The closest settlement was about ninety miles down the coast on Port Curtis, where the town of Gladstone, after an abortive beginning as a convict colony, had by the late 1850s become a port for squatters in the interior. It was being fostered by the New South Wales Government because Separation was coming and the people of Sydney did not want the new colony to have its capital too close to the rich Northern Rivers districts of the mother colony.

An able and energetic Government Resident, Police Magistrate and Crown Lands Commissioner had been found in Captain (later Sir) Maurice Charles O'Connell who, among his other efforts to develop the district, kept his eyes open for any traces of gold. On an expedition north in October and November 1857 he found about a pennyweight of gold near the Fitzroy River, and on his return dispatched W. C. Chapple (sometimes spelt Chapel) to follow up the find.

Chapple, a wiry, garrulous little Cornishman who had done some prospecting in New South Wales and later looked for copper and other minerals in the Gladstone district for O'Connell, was, as things turned out, not exactly the best man for the job.

He found a fairly rich deposit of gold on Canoona Station at a spot about seven miles from the Fitzroy River and thirty miles or so from Parker's shanty. Without waiting to work much of the ground, he headed south in great excitement to report to O'Connell at Gladstone.

On the way he camped at Gracemere Station and told the whole story of his find to some Germans the Archers had working for them. He also mentioned it to a couple of other men he met on the track.

He arrived back at Gladstone about the end of July 1858 and reported to O'Connell, supporting his story with some samples of good alluvial gold and a few fair-sized nuggets.

Then he took the rest of his gold to Captain Philip Hardy's Gladstone Hotel to celebrate. Before he had bought his third round of drinks the word had gone round and almost every man in Gladstone was in the bar demanding to know where the gold had come from.

Somewhat belatedly Chapple became cagey about the location, but he did produce a nugget weighing about twelve pennyweight and announce, "There's a ton of it there; a man doesn't even need to dig for it."

O'Connell took a cautious approach to the report and made arrangements to send the Government Resident Surgeon at Gladstone, Dr Robertson, who had been on the diggings in California, to check on the find. But by then everybody in town had heard of it and believed it to be the richest ever made on the Australian continent.

Publican Hardy, of the Gladstone Hotel, a former sea captain who had done some prospecting in New South Wales, wasted no time. Before the day was out he had raised about seventy pounds and organized a party of eight mounted men with bullock dray, rations and tools to go north to Canoona and try their luck.

Included in the party were Hardy himself, Hardy's black boy Pickwick, another hotel owner named Ellis, and Chapple who had been persuaded that he had now discharged his obligation to O'Connell and was entitled to do something for himself.

They reached Parker's Bush Inn on the second afternoon out, and the next morning arrived at Canoona to find Chapple's ground being worked by about a dozen Germans, some Chinese and a few others. Tense moments followed, with each party telling the other to get out and each determined not to budge. The apparent impasse was broken by Hardy's party moving in quietly and staking ground beside the others and no blood was shed.

Within an hour the new arrivals had worked seven and a half ounces of gold. After that the yield dropped off a good deal, but in less than a fortnight they had washed about twenty-two ounces.

Then Constable Frederick Woods, of Gladstone, arrived with a message from the Chief Constable that if Hardy and Ellis did not return to Gladstone forthwith, their hotels would be closed down. The publicans reluctantly decided to return next morning, but by then Hardy's horse was missing and he had a tedious ride back on Pickwick's mount, Paddy, an ancient animal on which Hardy had not spent a great deal of money.

From Gladstone, Hardy wrote to his agents in Sydney saying he was onto good gold at Canoona and that he expected to do well from it.

His was not the only report. Dr Robertson had returned with some good-looking specimens, and on 13 August 1858 James Atherton, camped near Parker's shanty, wrote: "I have ridden over the new goldfield on the north-west side of the Fitzroy. I saw ten ounces of the gold and a nugget weighing two and a quarter ounces. I believe it will turn out a very rich field. There are about a hundred men on the diggings."

Captain Parkins, of the schooner *Coquette*, said that on 24 August he had seen fine specimens at Captain O'Connell's residence, and four pounds of gold in the possession of a Chinaman. It was reported that a nugget weighing nearly three ounces had been found no more than five inches below the surface.

By 25 August 1858 it was reported, "Gladstone is deserted and nothing left but women and cradles." Everyone within reach rushed to the field. There was some fighting for ground and the Chinese were pushed out. Constable Woods reported: "Port Curtis is in a miserable, dilapidated state and all the women are going to the diggings. Several Jews are opening stores on the field. The Chinese took up arms in revolt but were disarmed and had to submit. The diggers are a poor lot, not like those of Victoria."

Hardy remained in Gladstone no longer than he had to, and after making arrangements about the lisense of his hotel, headed back to Canoona where he staked another claim. Though the field by this time was becoming crowded, he struck it lucky again and within a few days washed out thirty-three ounces of gold.

The ground at Canoona was shallow, with alluvial gold lying sometimes no more than a few inches beneath the surface on a bed of serpentine rock where it had been deposited by some ancient river. Near the surface particularly, the gold was stained black by other minerals. Many nuggets were completely black and distinguishable as gold only by their weight. Diggers called these nuggets "black gold" and some of the inexperienced, cradling their dirt, threw the nuggets out with the stones. The bedrock was so close to the surface in places that a man could prospect the whole of his claim in a day, sometimes in an hour. It was like a lucky dip. The digger knew almost at once if he had anything or not. If there was gold there he could work it out quickly; if not, there was no point in wasting time. In the later stages of the rush some diggers took off the top layer of the serpentine to get gold that had slipped into the cracks, but once the rock surface was cleaned the gold was finished.

Many grog shanties opened on the field, and tempers being short, there was a good deal of brawling and some bitter fighting. Life in general was rugged, even for a new goldfield.

About eighty ounces of Canoona gold reached Sydney early in September 1858, and without anybody waiting to see if it would keep coming the steamers *Yarra Yarra* and *Pirate* and nine sailing vessels were laid on to take miners north. By 21 September, 1,500 men had left Sydney for the rush.

Nine vessels were laid on at Melbourne, and from Sandhurst, from Castlemaine, from Ballarat, from every goldfield in the colony expectant diggers came trooping in to take ship to Canoona. On nothing more tangible than rumours that had been exaggerated over and over again in the telling, men left claims that were bringing them good money, rolled their swags and trudged in to ship north.

Victorian newspapers advised waiting for more definite information, and drew gloomy pictures of the north as hot, unhealthy country whose only proved products were small sweet potatoes and large snakes. The diggers paid not the slightest attention. Twenty-five sailing vessels and three more steamers were laid on at Melbourne for the Fitzroy River; the goldseekers crowded aboard them.

Further north, the whole country was in the grip of gold fever. Hamilton Ramsay, who had travelled overland from Canoona, arrived in Brisbane on 27 September with a hundred ounces of gold. Next day the editor of the *Brisbane Free Press* wrote in a leader: "We had one hundred ounces of Canoona gold on our office table last night, pure, good, nuggety gold." In the face of such reports warnings went unheeded.

With diggers streaming north into the rush area, which was still part of New South Wales, Captain O'Connell was gazetted as a Gold Commissioner on 17 September 1858. Ten days later he reported that while he still had faith in the field he viewed with some alarm the unusual numbers said to be on their way to it.

The tiny settlement around Parker's shanty was proclaimed the township of Rockhampton, and an Assistant Gold Commissioner, a Sub-Collector of Customs, a Landing Waiter and Tide Surveyor, and a small police force were sent to take up duties there. During one week in September about fifteen hundred diggers arrived on their way to Canoona. The ships brought them about forty miles up the river to the spot where a rock barrier stopped further progress, and landed them at Parker's shanty on the south bank. There they pitched their tents for the first night.

Stores and more shanties rose quickly, and a canvas town soon grew up. A couple of banks opened branches in tents where the managers slept on their money with pistols under their pillows. Streets were pegged out, and the bush was gradually cleared to provide building material. Demand for rations forced the price of flour up to forty pounds a ton.

By the time the men from the south arrived at the field there was little ground left that was worth pegging. The whole field was only a few acres in extent, and already most of it had been scraped down to the bare rock. Many who had come a thousand miles or more remained no more than an hour. Few stayed more than a couple of days before taking the weary track back to the south, to meet on the way hordes of still hopeful diggers coming up.

Hardy was stopped on the track by a party of disillusioned diggers on their way back to Sydney.

"Your name Hardy?" asked one.

"It is," he told them.

The man took a length of rope out of a wheelbarrow he had been pushing and pulled it between his hands as though testing its strength. "How would you like to have your neck stretched?" he asked.

Hardy looked around and saw he was surrounded by the rest of the diggers. He said nothing. The man with the rope went on.

"All the way from Sydney we've come, and look what we've got for our trouble—nothing. Is that what you call good gold? We'll likely die and rot in this blasted country because of the likes of you. But before we do . . ." He paused. With a sailor's skill he had been making a noose in the rope as he spoke. Then he went on: "Now there's a likely looking limb on that tree there, maties."

Hardy slid his hand into his coat pocket and dragged out a revolver. Calmly and deliberately, he took aim at the head of the self-appointed hangman. "How would you like to have an ounce of lead in your thick skull?" he asked.

The man caught the cold gleam in Hardy's eyes, dropped his rope and ran, with the rest of the would-be lynching party hot on his heels. Hardy pocketed the revolver and urged his horse to a gallop, thankful that none of them had called his bluff. He had fired his last cartridge long before leaving the goldfield.

Captain O'Connell also had a skirmish or two with disgruntled diggers, but without serious damage being done on either side, thanks mainly to the Commissioner's tact.

At the bark shanty village of Rockhampton, ships that had landed hordes of eager diggers were rushed by others in an even greater hurry to get out of the place. Many had no money and brawls were frequent.

Saddler John Scanlan joined the rush and arrived about the middle of September 1858. He found a great crowd of people camped in and around the town. Tents filled a square mile or more of the countryside, and frame houses covered with canvas formed several streets. Grog shanties and billiard saloons were all over the place and, in spite of the reported failure of the field, doing a roaring business. Streets were crowed with people talking and arguing about getting back south. Constables with bayonets on their carbines did the best they could to keep people moving.

Scanlan and some mates, humping their blueys, crossed the river a little above Yaamba in a boat and reached Canoona goldfield in just over two days. They met dozens of men on the road who told them the field was a duffer, but having come this far they were determined to go on and see for themselves. When they arrived they saw very little gold and not many who were even looking for it. The field seemed to be full of men with nothing to do. They staked a claim and for three weeks worked hard to get barely enough gold to pay for tucker. Then they gave up and headed back to Rockhampton.

Even then the town was still crowded and there were nearly as many men on their way up to the field as there were on their way back. South-bound vessels were all searched at Keppel Bay, at the mouth of the river, for stowaways, and those not able to pay their fare were put over the side.

Another who joined in the rush was Captain A. E. Sykes, who left his ship in Sydney to take passage north on the *Grand Trianon*, which brought seven hundred passengers. The vessel anchored off the mouth of the river and signalled for a pilot, but none came. Keppel Bay was crowded with vessels of all rigs and builds, including two ship's lifeboats which had been sailed and rowed all the way from Sydney.

The *Grand Trianon* had to wait in the bay for two days. While they were there a south-bound vessel, the *Jane*, passed them packed to the rails with returning passengers. One of these men cupped his hands to his mouth and yelled, "Go back, go back; it's a duffer; men are living on grass."

Straight away there was panic among the *Grand Trianon*'s seven hundred, who milled around demanding to be taken back to Sydney at once. The argument was still going on when the pilot finally arrived to take them up the river. Only 176 diggers, including Sykes, agreed to leave the ship; the rest remained, determined to make sure they kept their berths for the return trip.

The same kind of thing was happening on every other ship, but still they continued to arrive every day packed with men expecting to find gold. Rockhampton had no telegraphic link with the south, and word of the rush's failure had to wait for the first shipload of disillusioned men to reach home.

It was estimated that between fifteen thousand and twenty-five thousand men came from the south for the Canoona gold rush, and about half of them either never got off the ships or, if they did, never went further afield than Rockhampton. The Government of Victoria spent ten thousand pounds to bring its destitute diggers home. Not nearly so many had come from New South Wales, but the people collected about seventeen hundred pounds to help them.

Some of the more sturdy shouldered their swags and set off for the south overland. Others found work on the Archers' station.

Though the Canoona rush was disastrous to many who took part in it and made experienced southern diggers wary of northern gold for years afterwards, the field itself was not a failure. Within its limitations it was a good one, yielding many thousands of ounces of gold quickly with very little work. It also led to the rapid development of a valuable area which otherwise might have been neglected for years.

Chapple, the talkative digger who had started it all, was one of the first men the disillusioned diggers had turned on. Insulted, threatened, assaulted and burned in effigy, the little man was forced to abandon his claim and get off the field. Some time earlier he had found traces of gold on Alma Station, on the Calliope River south-west of Gladstone, and he went back there to scratch a living out of it.

One night, while suffering from a bout of fever, he fastened the door of his hut with only his dog inside with him for company and lay down on his bunk. Chapple died, and by the time his body was found by a passing stockman attracted by the frantic howling of the dog, the starving animal had partly devoured it.

Chapter Three:
The Battle of Crocodile Creek

Because of the failure of the Canoona rush, a large proportion of the Rockhampton district's early settlers were old diggers. The diehard prospectors spread out over the surrounding country still looking for gold. Those who took jobs on stations kept their eyes constantly skinned for indications of it.

As a result, the black nuggets of Canoona came to light again and again during the next ten years or so over widely scattered areas around Rockhampton. Field after field was opened and, though none were spectacular and most died out, a good deal of gold was won and there was considerable activity in some areas for many years. The gold that was found brought more men north and kept interest in prospecting alive.

It was ironical that O'Connell, who had sent Chapple looking for gold in the hope of developing Gladstone, achieved in the long run just the opposite. The new town of Rockhampton, about forty miles inland and on a river navigable to vessels of five hundred tons, was more convenient as a port for most of the stations than Gladstone, which was on the coast and cut off from the interior by rough country. Gladstone drifted into the doldrums and remained undeveloped for a hundred years.

Among diggers who came from Sydney looking for gold at Canoona was a sailor David Williams. After the gold cut out he prospected around with a former shipmate known as Bonnie Doon Bill. They got onto good gold and quietly settled down to washing out as much of it as they could in a hurry.

One evening they heard cooeeing from an adjoining hill and had only just time to cover up signs of their success before a stranger stumbled into their camp. He said he had got lost on his way to Canoona and had been attracted by the smoke of their fire. He would make it worth their while to show him the way to the diggings.

More interested in getting rid of him than in anything else, they decided that Bonnie Doon Bill should show him.

Bill took the man to Canoona, accepted a few brandies as a token of the digger's gratitude, and told everybody in the shanty that he and Dave Williams were on good gold of their own. The disgruntled diggers of Canoona rushed to join them, the find was worked out in a few days, and nothing was left but a collection of empty huts and some holes in the ground which they named Bonnie Doon after the man who had spread the news. Bonnie Doon Bill went back to the sea.

In 1861 a few prospectors were getting gold from a gully running into Sandy Creek in the Peak Downs district nearly two hundred miles west of Rockhampton. Other strikes were made in the surrounding country, copper was found nearby, and by 1863 Clermont, on Sandy Creek, and Copperfield, a few miles to the south, were thriving towns. A gold escort of mounted troopers ran regular consignments to Rockhampton.

Towards the end of 1863 word leaked out that two prospectors were doing well a mile or so north of Cowarral Station, about twelve miles north-east of Rockhampton. A mob went out and found four prospectors working hard but not inclined to talk. The mob prospected around, found no gold but a few stray colours, and demanded that the prospectors show them where the good ground was.

The prospectors told them to go to hell, and in the brawl that followed had their tent burnt, their tools smashed, and themselves beaten up. Nobody else found much gold there.

About the time gold was found in the Peak Downs district a man named Alexander Craig, on his way to set up a hotel on the Rockhampton-Peak Downs track, met on the road an agreeable young man who said his name was Frank Christie. With him was a very attractive young woman he introduced as his wife. Craig took them into partnership in a hotel which they set up at Apis Creek about a hundred miles north of Rockhampton.

Christie was in his early thirties, dark complexioned with thick brown hair and moustache; his wife was slight, fair and about twenty-five. Both went out of their way to help gold diggers on the road and were soon very popular. Miners often left their gold with them for safekeeping, and on a number of occasions Christie did duty as gold escort.

The district was electrified in March 1864 to hear that Christie had been arrested and accused of being Frank Gardiner, the bushranger who had robbed the Forbes gold escort in New South Wales back in 1862.

Gardiner had been betrayed to the police by a man who knew him in New South Wales in 1860 in the days of the Lambing Flat gold-rush when Gardiner, who at that time had already served a sentence for horse stealing, was butchering stolen cattle and selling the meat to the diggers. After that Gardiner had done some bushranging, fallen in love with Katie Brown, a stockman's wife, and left New South Wales with her to start a new life in Queensland. The New South Wales Government had a price of a thousand pounds on his head.

At indignant meetings of diggers who had been helped by Christie there was talk of raiding the Rockhampton jail and getting him out. There was also talk of lynching the man who had given him away.

The police lost no time in getting their prisoner onto the first ship for Sydney. There he was convicted on a number of counts arising out of his earlier exploits in New South Wales. He served ten years in jail and was then released on condition that he left the country.

Katie, who had stood by him during the trials and done everything she could to raise money for his defence, had by then exhausted her resources and committed suicide. Gardiner, the bushranger who tried to go straight, went to San Francisco where for many years he ran a saloon. He is believed to have been killed in a brawl.

Early in August 1864 digger G. P. Pilkington, who for some months had been selling gold that was supposed to have come from Peak Downs, came into Rockhampton and took out a summons against his mate, Joseph Moss, for assault. Sergeant McMahon was sent out to serve it and was taken by Pilkington, not to Peak Downs, but about twenty miles or so south-west to Stony Creek where he found a party of men who had already worked about five hundred yards of a gully and apparently done fairly well out of it. A small rush followed and the gold was soon gone.

From Stony Creek the diggers spread out over surrounding creeks and gullies and in June 1866 found good, payable gold further east in Crocodile Creek and surrounding gullies.

By early August there were about a thousand men on the field and more were still pouring in. The end of the year saw a shanty town on the site and more than three thousand looking for gold. About a thousand of them were Chinese, and there were also several hundred New Chums from Brisbane who knew nothing at all about getting gold and had not expected to do any more work than pick it up off the ground.

Work on Crocodile Creek was, in fact, exceptionally heavy. There were huge boulders that had to be rolled out of the way to get at the gravel, and much of the work had to be done in water and mud. Men who were not prepared to do this sort of digging said the Chinese had taken all the best claims, and old goldfield resentments against the Chinese began to flare up again.

Things came to a head early in January 1867 when a party of New Chums tried to jump a Chinese claim and in the scuffle that followed one of them was chopped across the skull with a tomahawk. He staggered off, hands to his head, blood pouring between his fingers, and bellowing for help.

Yells of "Roll up; roll up" echoed around the field; diggers erupted from their shafts like bull ants and came pounding to the rescue, waving picks, shovels, axes or anything else that was handy.

The Chinese dropped their tools and bolted for Chinatown where they hoped to find shelter in their buildings. Stones flew like hail as the diggers closed in. Huts on the outer fringes fell first. Doors were battered down, the occupants kicked, punched and driven out of them, interiors wrecked and plundered, and the buildings set on fire.

One that escaped was a shanty run by Mrs Ah Sing, wife of one of the Chinese diggers. As the attackers rushed the place she had lined up ready for them every glass in the house, each with a good stiff nip of brandy in it. Every man, as he came charging in, had a glass thrust into his hand. The diggers downed their free drinks and rushed on, leaving the shanty untouched.

About thirty Chinese tents and huts were burnt that day, and also several small stores. Many Chinese were manhandled and hurt, but none were killed. Some of the fleeing Chinese ran into the huts of European diggers and were sheltered there. Shouting crowds of rioters gathered outside but melted away at the sight of a rifle or six-shot revolver held in the hand of a determined-looking digger or his wife. As the first wave of the attack receded the Chinese rallied at a shanty about three miles out of town and sent a messenger on horseback to the police at Rockhampton.

Early next morning Gold Commissioner John Jardine, Sub-Inspector Elliott and about seventeen mounted police left Rockhampton for Crocodile Creek. As they neared the field they passed a steady stream of retreating Chinese who cheered them on their way with cries of, "Soon be all right me come back, eh?"

By the time they reached Crocodile Creek the place had quietened down. The white diggers had found out what started it all and, claim jumpers not being popular on any goldfield, had come round to the side of the Chinese, many of whom were already back at work on their claims. The number of white diggers who could not deny having taken part in the attack came down to a basic hundred or so. Some of them had gone bush when they heard the police were coming, the rest were thinking up alibis as they drank brandy in the shanties.

On inquiry it came out that the riot had been caused by less than fifty New Chums. Jardine and Elliott rounded up ten of the most likely suspects to take back to Rockhampton for trial. The men applied for bail and were refused. When they came before the magistrates two of them established alibis, and two more were discharged on the grounds of insufficient evidence. Six were committed for trial on charges of riotous behaviour.

In March, after they had already been in jail for a couple of months, they came up for trial. It quickly became obvious that the prosecution's main problem was identification. To the Chinese, all Europeans looked alike. Two men were acquitted. The remaining four were convicted of having committed an affray and sentenced to nine months each in the Brisbane jail.

Inconclusive though the result may have seemed, Jardine's prompt action discouraged, for a while at least, confrontations between Europeans and Chinese on northern goldfields.

The Crocodile Creek goldfield was worked out fairly quickly, but it produced about a hundred thousand ounces of gold and, in conjunction with the many smaller fields, played a part in keeping prospecting alive at a time when the new colony of Queensland badly needed it.

Nobody realized then that not far away, over the hills to the south, still part of a barely explored cattle run, was the hill that was one day to become Mount Morgan, one of Australia's great gold producers.

The most surprising feature of all this early prospecting, however, was that while men sweated at the great boulders of Crocodile Creek, away to the south on the Mary River several more men had handled the rich gold of the stinging-tree country and tossed it aside as of no interest. One of them was a bullocky following Bidwill's marked tree line. J. J. T. Barnett later wrote:

> In 1865 a start was made from the Mary River near Owanyilla by three teams of bullocks, the first being driven by the owner, W. C. Giles, the second by a Yankee named Mick Stanton. The third was in charge of a very capable bullock puncher and "white" blackboy, Johnny Walker, son of an old warrior, Peter, King of Tiaro. The horses and spare bullocks were in charge of Johnny Walker's gin, Maggie, and I went along to get Colonial experience, and got it too.
>
> We crawled along the track to Gutchy [Gootchie] where we saw old Andrew Puller and his black Maria. We went to Curra Station, seeing none on the way until

we met Mr R. J. Denman coming down the river driving a bullock team heavily laden with log cedar from the heads of the Mary River.

Twelve miles beyond Curra we came to very rugged country and camped on a gully which the track to Traveston crossed.

Having let go the tired bullocks, Giles took towel and soap and went to the gully to wash. He shortly returned, bringing in his hand a piece of quartz about the size of a hen's egg and sticking to it, lumps of bright yellow metal. This metal Giles—who had been on the Victorian goldfields in the fifties—said was gold, and no mistake about it.

Giles said he would form a prospecting party, but he forgot it all. In those days I knew nothing of gold mining. Had it grown on trees I might have plucked it, but dig it up like potatoes, I would not.

Chapter Four:
The Loner from Calliope

If ever a colony needed a big gold strike, it was Queensland in the 1860s. When separated from New South Wales in December 1859, it had 670,500 square miles of territory, most of it unoccupied and much of it unknown. It had about 23,500 people, and no money. Its early years were a battle to build population by immigration and get developmental work done without finance.

One of the larger projects being pushed through was the construction of a railway link between Ipswich and the Darling Downs which at the time produced a large part of the colony's wool—the mainstay of its existence. Ipswich, forty miles up the winding Brisbane River and its tributary the Bremer, was the main town, Brisbane being little more than a dilapidated former jail, with gum-trees encroaching on the dirt tracks of its outer streets.

As the 1860s advanced, prices for wool fell, drought hampered the settlement programme, and immigrants could not get work. The crash came early in 1866 when two of the main banks—Agra and Masterman's Bank of London, and the newly established Bank of Queensland—both closed their doors. Public works stopped, among them the construction of the Ipswich–Darling Downs railway.

Sacked navvies, whose wages had not been paid, filled Brisbane streets. There was talk of burning down Parliament House and breaking into shops to take food there was no money to buy. Field guns were set up in front of Government House, and shopkeepers were advised to batten their windows. There were riots and clashes between unemployed and police.

From the northern fields a trickle of gold flowed into the colony's Treasury. To stimulate prospecting, the Government, in January 1867, offered a series of rewards for the discovery of new fields, no nearer than twenty miles to an existing field and capable of supporting for six months a population of not less than three thousand persons. Many of the unemployed rolled swags, and before long every man in the bush had his eyes open for gold and his ears cocked for any rumour of it.

Though death from a spear between the ribs or a nulla-nulla across the skull was still a distinct possibility, the chances of it were not nearly as strong as in the days before Canoona. The coming of the sheep and cattle men had brought the inevitable clashes—driving off of Aborigines, retaliatory spearing of cattle and men, massive punitive raids and Native Police forays. Numbers of Aborigines were drastically reduced and those who remained thought twice about making an attack that might result in the destruction of the rest of their tribes.

There were rumours of gold in the air early in 1867 when Zachariah Skyring, a well-known Brisbane cattle and sheep dealer, went up to Nanango, about eighty miles to the north-west, to look at a big draft of sheep. Nanango at the time consisted of Bright's Burnett Inn—where Skyring put up for the night—and a few bark and slab huts built for the convenience of men from the surrounding stations.

Among Bright's customers that evening was an old digger from Victoria who was then working as a shepherd on Nanango Station. Skyring, who was always looking out for information about sheep, spent a good part of the evening talking and drinking with him.

As the night wore on, the Victorian began dropping hints about there being more valuable things than sheep around Nanango, and at last he asked Skyring to come down to his hut at the back of the inn. Arriving there, he opened the door and motioned Skyring to go in, then looked all around outside before following and closing the door after him. He lit a slush lamp that was standing on a rough bench and faced the dealer.

"Now," he said, "you're going out to the station to buy that flock of fats I've been looking after for the past two years, right?"

"Something like that," agreed Skyring, who was not sure what was coming next.

"Well, don't," said the shepherd.

"Why not?"

The man, who had drunk fairly heavily during the evening, steadied himself against the bench.

"Because," he said slowly, "there's more in Nanango than sheep. There's gold. I've found a goldfield at Nanango that licks Victoria into an old hat."

While he was talking, the old man had been unfastening the top of his trousers, and now he produced two bags, which had been tied round his waist, and handed them to the dealer. Each bag contained about a pound of good, alluvial gold. It was the first alluvial Skyring had ever seen, but it took no special skill to calculate that this sort of gold was a better proposition than sheep.

At first light next morning the two of them headed for the gold valley, which the shepherd said was about three miles out. They walked down to a small creek and there, from under a bush, the man produced a pick, a shovel and a dish. He led the way to a hole in the creek bank and dug out a dishful of dirt. He washed it and held the dish in the sun for Skyring to see. In the bottom gleamed about two pennyweight of gold.

All day they kept at it steadily, their only interruption coming from a party of puzzled, naked Aborigines who stood off watching them for a while before walking away.

As Skyring knew nothing about gold-mining it was agreed that he should go to Brisbane to secure their legal right to the claim. Two days later he spilled out on the desk of the Premier, R. G. W. Herbert, a bag of bright golden nuggets.

The Premier gazed at them dumbfounded, and then escorted Skyring to Government House and introduced him to the Governor, Sir George Bowen. Once again the gold was spilled out before astonished eyes. Lady Bowen, in answer to her husband's excited shout, joined them, and for a few hours Skyring found himself party to the Executive Councils of the colony which at that time was teetering on the brink of bankruptcy.

Next morning Skyring, legal title to the claim secured, sold his gold to Flavelle Brothers, jewellers, in Queen Street, and headed back for Nanango equipped with everything a prospector could possibly need. Behind him he left an excited crowd extending half-way across the wagon-rutted street in front of Flavelle Brothers, in whose window had been placed a silver dish piled high with gleaming nuggets and labelled "From Nanango Goldfield".

In the crowd that had gathered there were three diggers down from the Calliope goldfield. Word spread fast, and within a few days not one of the four hundred men who had been scratching a living out of Calliope remained on the field.

Among the diggers who came down to try their luck at Nanango was one man who had always stood out from the others. Diggers who were there remembered him as a smallish man in his early thirties, who camped alone on a solitary hill and never associated with anybody else at all. His name was Jim Nash, and that was about all any of them knew about him.

The old prospector from Victoria had been over-enthusiastic about Nanango. It came nowhere near any of the big Victorian goldfields, and under the busy picks of the diggers who flocked in from all over the colony soon showed signs of being worked out.

One morning soon after it had become obvious that the field was finished, diggers getting their breakfasts noticed that the loner from Calliope had gone. Nobody cared much what had happened to him, but on a goldfield men were curious about a thing like that. If a man did not talk much, you always felt he must know something that you should know yourself.

About a week after he had finished, the overseer from nearby Yabba Station called in at the Nanango field and mentioned that he had met one of the diggers, heavily swagged and accompanied by his dog, apparently making for

Durundur Station, away to the south. The overseer did not give him much chance of getting through because there was a big tribal fight in progress between the inland and the coastal blacks, and he felt pretty sure that if one side did not get the lone digger the other would.

The next thing the men at Nanango heard of the digger from Calliope was that he had made it to Durundur, and after a week's spell there had turned due north, following Bidwill's old track towards Maryborough. But the man who told them had got an impression of his own that the man had no intention of going to Maryborough at all, though he had no idea where else he could be going.

That was the last news they had of Jim Nash at Nanango. It was bad country that he had disappeared into, with a lot of thick scrub laced with spiny lawyer vines and Gympie-Gympie stinging-trees. As nothing more was heard of him it was presumed that he had perished as others had done before him.

Among all the diggers who speculated on the movements of the loner there was apparently none who had heard—or if he had heard, had paid any attention to—a yarn told by Mick Stanton, the man who had gone through with Giles's bullocks in 1865. Mick had turned up in Maryborough some months afterwards with several lumps of very rich quartz which he said he had found fifty or sixty miles up the Mary River. Mick was regarded as something of a nuisance around the town, and nobody took any notice of him.

It was, however, just the sort of story that a loner like Nash —if he ever heard it—would be likely to store away in the back of his mind for investigation some time when it suited him.

A few months after Nash disappeared, a little gold was discovered at Enoggera, on the western outskirts of Brisbane, and, Nanango being nearly worked out, many diggers left there to try their luck with the new find. About ten o'clock one morning a party of them walking down Queen Street, Brisbane, got the shock of their lives. One of them later wrote

> To our astonishment we saw a crowd of men gathered round Flavelle Brothers. On looking into the window we

saw two silver dishes heaped up with lumps of gold; to us they seemed as big as cricket balls. Some pieces, we were told, weighed up to nineteen ounces. The bulk was labelled: "From Cape River Diggings".

Captain Till, one of our party, and a late digger on the Cape, said, "It never came from there, I'll swear."

The captain went in to the manager for information, and was informed that the man who had sold the gold was staying at St. Patrick's Tavern up the street. We went there and saw a man standing alone against the lamp post. From the description, the Captain recognised him as the man who had sold the gold.

Directly he was spoken to the man disappeared like a shot from a gun and got into the back yard. My memory of him was good. I knew it was the man from Calliope. We decided to stalk him.

The landlord told us the man's board and bed were taken and paid for a few days ahead. A watch was kept for our quarry's return, but there was no sign of him that night.

Captain Cottier had left the wharf at 10 p.m. in the old paddlewheel steamer, *Yarra Yarra*, for Maryborough. Brisbane was only a township then and it did not take long to go through it, but without effect. Captain Till's ship, the *Policeman*, lay at the wharf from which the *Yarra Yarra* had gone. A loiterer on the wharf said he had seen a man answering to our description, earlier on the previous day, sitting on the wharf idling, and he remained there until Cottier went at 10 o'clock that night.

Next news the frustrated diggers had of the loner from Calliope was a report from Maryborough some time afterwards saying that a man named Nash had applied for a gold claim at a place called Gympie Creek between Widgee and Traveston stations, and that in support of his application he had produced great lumps of gold. The report went on that Nash had taken the mounted police and Surveyor Davidson to the spot, and that Nash had there dug out from the side of a gully in half an hour as much gold as a man could lift. More than a hundred ounces of gold was said to have been lodged by Nash in a Maryborough bank.

Chapter Five: Sitting on Gold

James Nash was born in Wiltshire, England, on 5 September 1834 and arrived in Sydney on 25 May 1858. He worked as a labourer for about six months and then went prospecting for about three years on the Turon goldfields, north of Bathurst, New South Wales. Word of a gold strike on the Snowy River sent him hiking south to Kiandra, where he worked for about five months with little success. He then tramped across to Twofold Bay and took a steamer back to Sydney.

About 1864 Nash moved north to Brisbane, did station work for a while, and then went looking for gold at Calliope. He had some success, but was nearly killed when his shaft fell in and buried him under six feet of earth. He was dug out half-suffocated and survived to join Skyring's Nanango rush. Soon satisfied that there was nothing worth staying for at Nanango, he rolled his swag and moved on about the middle of August 1867, taking with him nothing else but his prospector's dish, his pick and his dog.

Accounts of Nash's discovery of the new goldfield vary. He gave his own narrative to Aleck Nimey, author of the *Gympie Mining Handbook* of 1887, and wrote a more detailed account which was published in the *Gympie Times* of 15 October 1896.

According to these he left Nanango intending to head back to Calliope, prospecting all likely spots on the way. He travelled by Mount Stanley to Yabba Station, camping one night at the hut of a shepherd who had with him a young and very attractive Aboriginal woman and her two fine half-caste children. Unlike most shepherd's huts of that time, this one was spotlessly clean, and the woman, who spoke very good English, insisted on sending him on his way in the morning with a good supply of mutton.

Years later Nash's wife, Catherine, met the girl who gave her name as Peggy. She had by then left the white shepherd and was living with an Aboriginal by whom she had two attractive, graceful, jet black piccaninnies. She proudly told Mrs Nash: "I knew Jim before you did; when he came to my place he only had wallaby meat; I gave him mutton."

Still later, Catherine Nash met Peggy again. By this time she had left her Aboriginal husband and proudly anounced that she was now married to "a gentleman". This man was a fine young half-cast, son of a squatter and a native woman.

Nash, after leaving the hut of Peggy and the white shepherd, stayed a night at Yabba Station and moved on next day. The following night he camped out with two boys who were tailing cattle about twelve miles from Yabba.

The boys seeing his prospecting dish, told him there had been some gold found at a place three miles away called Bella Creek. After getting some promising colours in several places on the creek, Nash made a quick trip to Brisbane to spend the last of his money on a horse and rations and then returned to prospect the area thoroughly. It did not come up to his hopes and he pushed on to Imbil Station, where he camped for the night before going on to a meeting with Mr R. J. Denman, which was to be argued about for years.

Denman was the timber-getter whom Barnett mentioned having met just before he and Giles found and walked away from the Gympie gold in 1865, and Denman was to claim later that he put Nash onto the gold, and Nash cheated him out of his share of it. Nash denied this and said Denman's claims were all lies. Nash said:

> I left Imbil next morning, got to Denman's camp about 11 o'clock, saw a fire, boiled the billy and had dinner. Just as I had finished Bob Wannel came up; he was hauling timber for Denman. I asked him to tell me where to cross the Mary River and he told me he would send a boy with me after dinner. I told him I had taken dinner, but he wanted me to have a drink of tea for friendship's sake. After he had dinner we all went to

Denman's tent; he said he was an old digger from Victoria. I showed him the few specks from Bella Creek; he told me the Six Mile Creek would be a likely place for gold.

Mr Wannel sent a boy to show me the crossing; when I did cross I was searching for the track until the boy came looking for his bullocks. He put me on it and I reached Traveston Station that night. I left there next morning and got to the Six Mile Creek, but not liking the look of it, I did not try it at all, nor any other place until travelling down what is now Caledonian Hill.

It was about midday by then, and at the foot of the hill Nash dropped his swag near a gully that crossed the track to Maryborough. There was a waterhole nearby, and after lighting a fire and putting his billy on to boil, Nash—born prospector that he was—took his pick and dish to try a prospect.

He washed a dishful of dirt that lay in the bottom of the gully and got a clear tail of gold in his dish. He moved a little farther up the gully and, selecting his spot with more care from the face of the bedrock, took another dish of dirt and carried it to the water to wash. Before he was done washing he picked a small nugget out of the dish, and then another. There was more gold in the dish when he had finished. He stood up slowly and looked all around, studying the lie of the land and the rocks where they outcropped. His practised eye detected signs that could mean only one thing.

After he had finished his dinner Nash covered all traces of his fire and removed his swag up round a bend of the gully so his camp would not be seen by anybody coming along the track. Only when he was sure he had left no trace of his presence did he go on with his prospecting. He camped there that night—somewhere on the Caledonian Hill side of what later became known as Nash's Gully—knowing that he lay down surrounded by gold.

Next day Nash went a little higher up the gully and made a drive into the bank. He washed out an ounce and three pennyweight of gold and was on ground that seemed to be improving when he broke his hammer-headed miner's pick. Unable to work without it, there was nothing he could do but go on to Maryborough, sixty miles down-river, and exchange his gold for a new pick and rations.

Maryborough by this time was a town of about fifteen hundred people. It had grown up as a port for supplying the surrounding stations and shipping out wool, hides, and tallow. It also exported timber—mainly cedar for Victoria—and had lately become the centre of a growing sugar industry. But Maryborough had never been a gold town, and Nash had trouble trading his nuggets.

"I tried two banks and several stores but could not sell the gold," he said. "Times were so bad that they hardly knew what gold was like. At last I tried Mr W. Southerden [a well-known storekeeper of those days] a second time and he allowed me three pounds for it—one pound in money, the rest in tools and rations."

Gold at that time was selling in Brisbane for about three pounds an ounce.

Nash wrote to his brother John asking him to join him, and headed back to his find, arriving about ten or twelve days after he had left it. There was nothing of Bonnie Doon Bill about Jimmy Nash, and he had said not a word to anyone that would give the least hint of where his gold had come from or the fact that his find was a rich one.

He was fortunate in that it was a lonely place, near a sharp bend of the Mary River, part of Widgee Station, with Traveston Station on the south and Curra (also called Currie) Station on the north. Some distance downstream from Nash's Gully a pretty little brook wound its way between casuarinas to join the river. Its clear water had made it a favourite camping spot for cedar-getters and station hands. Somebody had given it the undeserved name of Gympie Creek. It was about the only named natural feature in the locality at the time.

As soon as he got back Nash started digging at his old spot. The water became muddy, he moved further up—and found gold all around him. "While washing the first dish there I picked up gold beside me in small pieces. The first dish showed gold freely—in fact it was gold all around me, and I did not go back to the first place after that. The actual place where I got the first gold was in a little branch gully in the bottom of Caledonian Hill. This gully has long ago been filled up."

In constant fear of being seen by somebody passing along the track to Maryborough, Nash worked as quickly as he could, careful always that no smoke from his fire in the daytime or glow of coals after dark should betray his presence.

Often, as he was working, Aborigines, some of whom had picked up a smattering of English while working on stations, would gather round and ask him: "What for white feller him look out?" Every time Nash would reply: "Looking for stuff same as other white man at Kilkivan." At that time copper was being mined at Kilkivan, about thirty miles to the west, so if the Aborigines had reported his activities, none would have realized he was on gold.

In six days digging Nash got seventy-five ounces of gold. He then decided to return to Maryborough, "there being no longer any doubt in my mind about it being a genuine goldfield". He was, in fact, expecting his brother John to be waiting for him there so they could report the find together. He covered all signs of his work and headed down-river.

On the way he camped for the night at Curra Station, and in the morning he helped to bury a blackfellow who had died there. The strain of keeping an eye on his gold without letting anybody suspect that he had it must have been considerable.

The journey to Maryborough took five days. When he arrived at the Sydney Hotel, where his brother was to meet him, he found that John had not arrived. Not wanting to report his find and start a rush without telling his brother about it, but being in need of money for stores and unable to sell his gold locally, Nash used a nugget to pay his steamer fare to Brisbane. On board the ship he became friendly with a young sailor named Billy Malcolm. When they reached Brisbane they both booked in at the St. Patrick's Hotel.

Nash took his gold to Flavelle Brothers and sold it, under the eyes of several curious observers, for two hundred pounds.

"The Honourable William Henry Walsh [Member for Maryborough] was in the shop as I came in, and watched me sell the gold, and there was a man watching from the street whom I knew was a digger. Mr Walsh asked me where I had got it and I said, 'Oh, up north'."

For the rest of his stay in Brisbane Nash was tailed by one digger or another wherever he went. To those who spoke to him he did his best to give the impression that his gold had been collected over a long period. He said he had been digging on the Cape River. Not even to Billy Malcolm, whom he had invited to join him, did Nash disclose the locality of his find. He bought a horse, a dray and a gold-washing cradle, and the two of them took the first steamer they could get back to Maryborough.

Brother John still had not arrived when they got there, so, leaving a message for him at the Sydney Hotel, Nash bought a tarpaulin, corn, chaff, and rations for nearly six months and headed back to the field with Malcolm. Had he set out so equipped from Rockhampton he would no doubt have had every digger in town at his heels, but the people of Maryborough were not used to thinking about gold, and Nash apparently succeeded in making them believe he was setting out on a very much longer journey than was actually the case. As it was, the sixty-mile trip took them about nine days, because the old bullock track was closer to the coast and the dray had to make its own track through untouched country.

Back on the field Nash was on tenterhooks about not having reported the find. After the sale of the gold in Brisbane it was only a matter of time before someone more persistent than most managed to track him down. Any bushman could follow a man with a dray if it seemed worth his while, and even though a man might hide himself and his equipment, disturbed ground along a creek bank or a patch of muddy water in a clear creek was indication enough of where prospecting was going on.

Any digger finding payable gold and reporting his discovery was entitled to a special prospector's claim varying in size according to its closeness to other diggings. He would also be entitled to the Government reward as the finder of a new goldfield. If, on the other hand, he kept it secret and somebody else found it, he lost both the reward and his claim to extra ground.

Sitting on Gold | 37

Day after day Nash and Malcolm worked as hard and as fast as they could. Several times a stockman from Widgee Station rode past and they took cover and remained quiet. One day he came on them so suddenly there was no time to hide. "Getting any gold?" he asked.

Both prospectors put on their most hangdog expressions. No, they said, they were not doing much good. They had been all over the surrounding country and were hardly making tucker.

Once a stranger passed along the track and Nash's dog began barking. Nash sprang to grab him and keep him quiet while Malcolm did the best he could to cover the ground that had been dug without making too much noise about it. The man stopped, looked around him, and then, with a shrug of the shoulders, went on.

After a fortnight of it Nash could stand the strain no longer. He had waited for John long enough. He would report the find and so make sure of the discovery claim and reward money. Once again covering all signs of work as well as possible, and hiding the dray and other equipment in the bush, the two of them mounted the dray horse and headed for town.

Nash's own narrative says his brother John had meanwhile picked up his letter at the Sydney Hotel and that John passed them on his way to Gympie Creek without either of them seeing the other. On the other hand, newspaper reports and the recollections of a digger named Leishman were to the effect that both brothers were in Maryborough when the find was reported.

Whatever the truth of that may be, Nash went to the Maryborough police station and on 16 October 1867 reported to Sergeant Ware the discovery of gold at Gympie Creek. There was no Gold Commissioner in the town so the sergeant referred him to H. B. Sheridan, Sub-Collector of Customs and Police Magistrate.

Sheridan received them cordially, and arrangements were made to swear in the Land Commissioner, Mr William Davidson, as Acting Gold Commissioner and have him and Sergeant Ware accompany the prospectors back to the field and peg out their claims.

By this time Maryborough was humming like a hive. There was a stampede to take out Miners' Rights and stock up with stores, and before the day was out there was hardly a pick, shovel or horse to be had in the town.

Back in Brisbane, however, Canoona still cast its shadow. Next day the *Brisbane Courier* cautiously printed a small paragraph from Maryborough stating: "Two brothers named Nash came to town yesterday and reported a find of gold weighing seventy-five ounces on the Mary River near Widgee's Crossing."

Chapter Six:
First on the Field

In Maryborough at the time Nash arrived to report his find was one of those men who always seem to be in the right place at the right time.

Former seaman W. Leishman had come from England to Brisbane in 1857, had operated a string of punts on the river, and then, following the bank crashes, joined Tom Findlay and another man in a venture to mine coal on the Burrum River, eighteen miles north of Maryborough. He took with him his wife, two young daughters, and son William. The only other woman in that part of the country at the time was a Mrs Howard, who could shoot like a man, speak the local dialect like an Aboriginal, and after whose husband a town that grew up on the Burrum was named.

With a good quantity of coal at grass, Leishman went to Maryborough to try to sell it. There, who should he run into but his old shipmate Billy Malcolm. Malcolm told him that he had teamed up with Jim Nash, that Nash had discovered gold near Gympie Creek, about sixty miles up the Mary River, and that they were in town to report the discovery. Billy introduced Leishman to Nash, who showed him some of the gold he had brought to Maryborough.

Leishman lost no time in deciding the Burrum coal could remain at grass as far as he was concerned. As Nash had to wait in town for some official arrangements to be made, it was agreed that Billy Malcolm, Leishman and Nash's brother John should go straight back to Gympie Creek and pick out their claims.

Malcolm and John Nash had already bought horses. Leishman had enough money on him for a horse, but not for a saddle and bridle as well. A man lent him a bridle and somebody else gave him a surcingle and stirrup leathers. He got some corn sacks and made a saddle with them. He sent word to his wife that he had gone to Gympie Creek, and the three men headed up-river.

They camped overnight in the bush and rode on to Nash's Gully next morning. That night Jim Nash arrived with William Davidson, Acting Gold Commissioner, and also Sergeant Ware of the police, John Cartwright, Charlie Brown, Maurice Walsh, William Walsh and a few others.

The following morning Nash had to wash a few pans of dirt to show the ground was payable and that he was entitled to the prospector's claim. Every dish showed not only gold, but small nuggets of it.

Davidson put the compass on Nash's claim and marked off an extra claim and a half at the bottom of it as prospector's reward claim. Then came John Nash with a forty-foot claim adjoining and then, on a gully running into the head of Nash's Gully, Billy Malcolm and Leishman. They called it Sailor's Gully, because both of them were ex-sailors.

During the day men began arriving from Maryborough, and the stream kept up without slackening for the next two days. Those already there could see them reach the top of the ridge overlooking the gullies in small parties. The real diggers among them would pause, run a quick eye over the lie of the land and the position of claims already pegged, and then, often dropping some of their heavy gear to be retreived later, come charging down the slope full tilt.

Those who were in parties split up, each man heading for the spot that looked most likely to him. Those on horses came at a gallop, leaving the foot-sloggers to swallow their dust. If a man saw someone he knew galloping past him on a horse he would call out, "Peg a claim alongside you for me, mate," and stagger on as well as he could to his own chosen spot.

Men who knew anything about it had their pegs already cut and sharpened in advance. They reined their horses on their chosen ground, grabbed pegs and hammer-headed pick from the saddle pack, and slammed the first peg into the ground. Each man was entitled to a claim forty feet square—fourteen yards was near enough for now—and each chose his next corner and paced it off. Often he ran into another hammering in his own peg half-way along.

"This is my ground; I was here first."

"Like bloody hell it is!"

Angry words flew—but no time to argue or another would come and take the ground while they disputed it. Quick compromises among old diggers—jostling aside of New Chums—a snarled threat or two—and the next peg was hammered home. Fourteen paces at right angles to it—another peg—and then the last—forty feet of ground pegged out and ready for the Gold Commissioner's inspection and approval. Time enough then for a man to tether his horse and look around him.

The place was pandemonium. Fights raged over disputed ground, horses plunged and stamped, dogs yapped and snarled, men shouted, cursed and yelled.

"Hey, young Jack, over here!"

"Peg me a claim alongside you, old-timer!"

"To hell with the Commissioner; my pegs are in the ground."

Some of the methodical ones were putting their ground in order, digging shallow trenches at right-angles at each corner of their claims to mark them more clearly, or shovelling out a rectangle of soil to mark the position of a future shaft. Others were content to sit on their claims and watch, maybe smoke a pipe. But no matter whether a man be thirsty or hungry, in need of stores or anything else, he did not, if he was wise, leave his claim until things had settled down.

Still roaming around the gullies, uncertain what to do next, were the New Chums. Some of them had never looked for gold in their lives, some would not recognize it if they saw it. Their tools ranged from picks and shovels to anything at all—hoes, spades, mattocks and crowbars. Many had come with no tools or rations or even money to buy them, many would leave the field next day without ever putting a pick in the ground or washing a dish of dirt.

There were even some old prospectors there who had made no attempt to drive in their pegs, and had abandoned all thought of digging because they considered it was too far to carry the wash dirt to the water. All the promising ground close to water had been taken.

Most of the old diggers always pegged ground for their mates as well as themselves. When there was a party working together members separated and each man pegged out enough claims for all of them. That gave them time to try as much ground as possible and see where the gold was. The best claims were then worked, the others abandoned. It was always easiest for a party to work together. Their combined claims gave them enough ground to work efficiently, division of labour made things easier for everybody, and a party of men were less likely to be worried by claim jumpers or gold thieves than one man on his own. Even Jimmy Nash realized from the start that this was going to be too big a rush for a loner.

There was no formal registration of alluvial claims; pegging out and holding a miner's right was all that was needed. A man had the exclusive right to his claim so long as he worked on it every day—Sunday excepted. If he left it for more than twenty-four hours it could be taken by somebody else, or "jumped". Some tried to jump claims whenever they saw them unoccupied, and this, together with another dodge of moving or replacing a man's pegs in the night time, were the main causes of goldfield disputes. Twenty-four hours was time enough for claim-jumping fights to flare up over the ridges around Nash's Gully.

Back on the Burrum River Mrs Leishman told Mrs Howard her husband had gone to Gympie Creek, wherever that was, and that she and the kids were going to follow him there. She got a lift in a spring cart to Maryborough and found the place in a turmoil.

On Saturday, 19 October, several who had followed Nash's party returned to Maryborough with 120 ounces of gold, including a 41-ounce nugget and several smaller ones. Within an hour of their arrival there was a stream of men on the track heading out for their share. All night the ferry kept going to take them across the river. By Sunday it was a stampede to get out of town. Farmers walked off their land. The sugar harvest came to a standstill. The boiling-down works closed. Ships were left without crews.

The schooner *Rose* was already on her way down-river when her crew got the news and dived over the side to a man. There was nothing the captain could do but let go the anchor and wait until an incoming ship lent him hands to get back up the river.

Among the early ones to get away was the formerly doubting storekeeper, Southerden. His notice appeared in the *Maryborough Chronicle* of 23 October 1867.

> OFF TO THE DIGGINGS
> W. Southerden informs his friends trying their fortune at the diggings that he will open a temporary store on the new field, where the necessities of life and the usual diggers' outfits will be kept for sale. Gold will be exchanged for goods, and as the accommodation will be temporary, parties must in all cases pay for goods on delivery. Arrangements are made for regular communication to and from the port.

Young Billy Leishman, who was not much over four years old at the time later described the family's trip to the new goldfield.

> The first team with provisions had left the day we arrived in Maryborough, but we caught the next team. We had a very bad trip, having to get off at every bad crossing. We practically caught up with the first team as they had to make crossings over the gullies, and we followed their tracks. The first team was just about two hours ahead of us.
>
> When our team got to the bottom of the hill [now the site of the Gympie Memorial Gates and Park] there were a few men putting up tents, and when they saw mother and us kiddies they cheered and hoorayed as we got off the dray.
>
> The men came and carried my sisters, and I toddled along with my mother. One of the men went on to where the first team was unloading. He asked if anyone knew where he could find a man named Leishman.
>
> Father was on the first dray shelling out provisions and the man told him that his wife and children were looking for him. He told the man to go to hell, and asked him if he was trying to pull his leg.

44 | Gympie Gold

Young Billy later recalled the family reunion out among the already pot-holed gullies, with stores stacked all around them, and grinning gold-diggers pressing close for a look at them. As Mrs Leishman came up to the main body of the men she and the children were cheered and welcomed on every hand. One of the men stepped forward and said: "Welcome missus; you are the first white woman on the field, and your kiddies are the first children."

The diggers regarded their arrival as a good omen for the permanence of the field and, leaving Leishman himself hardly anything to do, they rallied round with axes and tomahawks and before nightfall had erected a comfortable tent with bunks and other bush furnishings ready for the family to move into.

When the Nash brothers, Malcolm, and Leishman settled down to working their claims, Mrs Leishman cooked and acted as banker for the party. They had a box divided into four compartments, and each put his gold into his own compartment. The box was kept in a large, emigrant's sea-chest locked with a heavy padlock. There was no such thing as a bank on the field, and by the middle of November the chest was reputed to hold about forty pounds of gold.

The new goldfield was officially proclaimed on 30 October 1867 as the Upper Mary River Goldfield, though it continued to be referred to as Gympie Creek. Diggers on the field preferred to call the growing encampment of tents and bark shanties Nashville, after Jimmy Nash who had started it all, and Nashville it was for several months until officialdom ruled in favour of calling it Gympie.

Early reports from Nashville reflected in turn the elation of discovery and the fear of another Canoona. There were conflicts of interest between those who saw the goldfield making Maryborough a prosperous port, and those who faced ruin because of the unharvested sugar cane and closed boiling-down works.

A Maryborough man who apparently had an interest in getting his sugar-cane crushed wrote sourly to the *Queenslander*:

Maryborough since the goldrush has been something like Goldsmith's deserted village. Gold is still the all-pervading topic of discussion. Meet whom you may, no other topic can be thought of, but I think, notwithstanding the present attractions of this new pursuit, we shall soon resume the more reasonable and legitimate occupations.

I very much fear that the consequences already to many of our farmers will be, in a pecuniary sense, most serious. It seems a pity that this discovery of gold could not have been deferred for three months, for then few of the community would have been affected by it. But I have reason to believe that within a few days we shall have a return of the disappointed ones who will doubtless be glad to fall back into their usual and ordinary means of living.

In Brisbane the *Queenslander*'s editorial on 2 November also took a cautious line:

The excitement about the goldfield on the Upper Mary River is far greater than is warranted. Gold has been found in two or three gullies and a few men have been lucky in obtaining heavy prizes. The immediate consequence is the assembling together of an immense number of people from the surrounding districts, many of them having left steady employment in order to try their luck. As there are already more people there than can find room on the ground which is at present known to be auriferous a large majority must be unsuccessful.

But nothing was going to stop the rush. The next issue of the *Queenslander* carried a letter from the paper's Maryborough correspondent:

MARYBOROUGH, November 5: I should have written to you ere this, but I got the epidemic and rushed off to the goldfield. I was fortunate to get a claim in White's Gully and I certainly have no reason to complain. I tried a few dishes and got the colour in every one, and can guarantee an ounce and a half to the load of twenty buckets. . . .

I was obliged to come down to the District Court here on Monday last for jury service but am off back now Friday or Saturday.

Old diggers on the field were writing away to their mates. One veteran from California, Ballarat, and Gulgong wrote:

> Got thirteen ounces three pennyweight out of a little better than a load of dirt. Not so dusty for a week's work! So make tracks, Old Flick; don't let the grass grow under your feet. Don't be scared by the croakers. Such a rummy lot of codgers we've got here—counter-jumpers, street corner loafers, wharf idlers, cockatoo farmers and such like—who expected to break their shins over nuggets as big as pumpkins, are running about the diggings like cats on hot bricks looking for pieces of gold on the surface, and because they don't pick it up in hatfulls, are rushing back to their wives and mothers.

Even those on good gold had their problems, the worst of which was shortage of water. A newspaper correspondent reported:

> Prospecting with water two or three miles off is not likely to be prosecuted with any great vigour. There is a strange medley of every class and calling on the diggings, wending their way along the hills to water with bags, dishes, buckets, billies and even pint pots of dirt.

Queensland had no mining legislation at that time and was still operating—more or less—under the New South Wales regulations. One of these was that diggers, when on apparently payable gold, must hoist a red flag not less than a foot square near their working shaft. Failure to do so could mean forfeiture of the claim. There were red flags flying all over the field, but other prospectors had scattered out into the bush, and some were apparently doing well and keeping quiet about it. There were three men who came in from the scrub, exchanged forty ounces of gold for stores, and disappeared again. Some tried to track them, but soon lost the scent. Another party came in from the opposite direction, exchanged gold for rations, and vanished.

The first Chinese arrived on the field within a month of its opening. About thirty of them camped on the ridge above the township. Some of the Europeans nailed notices on trees calling on diggers to roll up and run them off the field, but there was little response and the Chinese stayed. The local Aborigines were no problem at all. The squatters' land wars had broken their spirit and those who were left seemed to resign themselves to the hordes of diggers who appeared suddenly in their midst.

The Nashville cemetery was started when the body of a man was found under a sheet of bark near the foot track leading out to a spot called the Two Mile. The Commissioner was informed and a court was convened, without furniture, under a shady tree. A crowd of several hundred gathered to watch. A doctor was found dressed in moleskins and working on a nearby claim. He examined the body, was duly sworn, and gave evidence that the deceased had been dead at least two hours. Surrounding diggers were called on to come forward if they could identify the dead man or throw any light on his death. None came.

"Now we come to the matter of burial," said the Commissioner.

There was some hesitation, and then a digger spoke up. "I'll put him away for a quid, Your Worship."

The Commissioner looked around the crowd. There were no other takers. "Done," he said.

There was still no graveyard, so the Commissioner marked off four acres of land about two miles from the diggings and ordered that four posts be put in the ground and a notice on a sheet of bark nailed up to indicate a reserve for burial purposes.

Two sheets of bark were brought to make a coffin, and the unknown gold-seeker was laid to rest.

The following week some diggers going to work at daylight found a man dead just outside Greathead's shanty. Once again a court was convened, and medical evidence was that the man had died of "constitutional weakness". This was corroborated by a digger who said the man had been his travelling mate up to Nashville and was very weak from the track. He added, "But I reckon if he'd only had the strength left to get inside for a drop of brandy, a kick from a mule wouldn't have killed him."

They put the body on a sheet of bark and carried it to the cemetery. Hundreds of diggers straggled along behind.

The Commissioner's Court was fairly busy during the early months. Cases were concerned mainly with boundary pegging and claim jumping, and from twenty to thirty hearings a day was not unusual. They were conducted with the minimum of formality.

The Gold Commissioner, accompanied by a mounted trooper, was constantly riding around the field, and when a dispute arose he generally saw it and came riding up to investigate; if not, he was sent for. All that was needed to start proceedings was for one of the parties to ask the Commissioner to adjudicate. Then the parties would gather under the nearest shady tree, the orderly would call out, "Now gentlemen, this court is open for the conduct of business," and the hearing would begin.

An old digger claimed proceedings were often interrupted by another argument, a thunderstorm, or some other cause, and that when this happened the Commissioner generally adjourned for an hour.

> The adjournment was nearly always made to the nearest shelter where refreshment was available and on resumption, the plaintiff and defendant would often be embracing each other and swearing eternal friendship. Or they might have decided to settle the matter in a more direct manner and when the court was resumed one might appear with the beginning of a black eye or some teeth missing. The other name would be called and there would be no answer, and judgment would be given by default. Only after the court had risen was it customary to tell the Commissioner that the losing party had been taken to his tent, maybe with a broken jaw or some other impediment to continuation of litigation.

After the first few weeks the rush began to slacken off. The first of the disappointed New Chums had arrived back in Maryborough and Brisbane and spread gloomy tales of their own lack of success. Newspapers had played down the rich finds from the start, fearing a rush would quickly clean out the field and leave thousands stranded. Diggers in New

South Wales and Victoria, wary of another Canoona debacle, waited for more definite proof of the field's permanence.

The result was a respite for the real diggers from the northern fields who now had a chance to look around and make sure they had the best ground before the old-timers came trooping up from New South Wales and Victoria.

Among others who got in early and did well was carrier John Palmer. When Nash reported his discovery Palmer was in Maryborough with his bullock team. He had brought down wool from a station on the Burnett and was loading stores to take back. When he saw the men streaming up-river to Gympie Creek he forgot about the station. There might be gold at Gympie Creek or there might not, but certain it was that with a crowd like that going to the field, a man with a wagonload of rations would be able to ask his own price for them. Palmer loaded up with flour, sugar, tea, and a few kegs of rum and headed up-river for the goldfield, taking with him his wife and young son Billy on top of the load.

There was already a rough track to the diggings, but no road for a bullock wagon. The cedar teams had not been through this part of the country and the few drays that had already gone had made their own ways wherever the country seemed clearest. Palmer had to chop through bush and scrub, work his way round steep hills, and explore every creek to find the best place for bullocks and a heavily laden wagon to cross.

He had been on the track three weeks and was within nine miles of the diggings when he found a very steep pinch in front of him and decided to make camp for the night. Next day he rode on to scout out the track and fell in with a party of glum-faced diggers humping their swags.

"What are you coming back for?" asked Palmer. "No gold?"

One of the diggers grunted in disgust. "It's lousy with gold," he said, "but there's no tucker; a man's got to get out or be starved out."

Palmer laughed with relief. "If that's all that's wrong you've got nothing to worry about," he told them. "I've got plenty of tucker."

He took them back to his camp and ladled out all the stores they could carry. He had no scales or measure, so he used a tin pint pot. He sold flour for a shilling a pint, tea and sugar for a shilling a half pint, rum for a shilling a glass, and so on. He made up the prices as he went along and the diggers were glad to pay them.

They said stores were so short on the diggings that he would soon sell out all that he had. In the face of this Palmer decided there was no reason why he should labour up the pinch and continue the tedious process of cutting a track to the field when his customers would be glad to come to him.

He built a permanent camp on the spot, unloaded his stores and, leaving his wife and young Billy in charge, headed back to Maryborough for more. In about three weeks he had returned with another load, and by then the first was all sold. The same thing happened a second time, and Palmer abandoned all idea of going on to the goldfield. He built a humpy of saplings and bark and gradually added to it until Palmer's Nine Mile Hotel became the best store and accommodation on the track to the field.

About the only farmer in that part of the country at the time of the rush was Mr A. McDonald, who grew maize on the banks of the Mary. Previously he had punted it down to Maryborough, where he had to pay fourpence a bag tax on it, before shipping it—if there was a boat available—for whatever freight the captain chose to demand, to Rockhampton; here it sold for about two shillings and sixpence a bushel, less commission. Once the rush started McDonald sold his maize, for whatever price he liked to ask, to the first batch of diggers to come along.

Another man who did well was the lessee of the ferry over the Mary River at Maryborough. Suddenly faced by a flood of diggers clamouring to get over the river, he kept it going day and night at charges ranging from sixteen to eighteen shillings for a bullock team down to two shillings for a man on a horse. Before the first flush of the rush was over he had netted more than two thousand pounds.

One of the rush's disgruntled men was Denman, the timber-getter who claimed he put Nash onto the gold and was cheated out of his share. His story, as told to Archibald

Meston who quoted it in his *Geographic History of Queensland*, was that in September 1867 Nash called at his camp and Denman told him to go to a likely spot he had seen and try it on the understanding that any success would be shared between them. He claimed the spot was the afterwards famous Nash's Gully. Meston said that Denman, who afterwards became Crown Lands Ranger at Maryborough, complained bitterly to him of Nash's breach of faith in not sharing his discovery.

Nash denied that Denman had any part in the discovery. "I never met him until I called at the camp when I could not find the crossing, and then did not have a meal with him, nor did I try the only place he told me would be likely to have gold in," said Nash. "Before I reported he passed one day and wished to know how I was doing; I told him there was a little gold there, and offered him the dish and pick to try his luck, but he said no, he had had bad luck in Victoria and he would not go in for digging again, he had his timber to look after. I did not see him again till after the rush."

A man who came back to see what he had missed was Barnett, the man who, two years before, had seen the gold and scorned to "dig it up like potatoes". He went back to the field and saw a claim being worked almost exactly on the spot where he and Giles had camped when they came through with the bullock team. Later again, he was told that the claim had yielded sixteen thousand pounds worth of alluvial gold.

Chapter Seven: Mother of Gold

It had been obvious to Nash from that first night when he sat against the boulder and surveyed the surrounding ridges that the gold he had found must have come from reefs that lined their crests. It had fallen out of them as they were weathered away and been washed down the slopes of the creeks and gullies. Some of it had been caught on rocky ledges, some had worked its way down, because of its weight, to the solid rock bed beneath a layer of lighter alluvial debris. Many of the nuggets had pieces of quartz still attached to them.

The story was plain to every experienced prospector who came to the field—the alluvial gold had come from the reefs; there must be more where it came from.

In the first rush the reefs were ignored in the scramble to get at the alluvial which lay near the surface or at the bottom of shallow shafts dug through rubble. Picking through the brown, slaty rock in which the quartz reefs were embedded would be harder work, and then the quartz would have to be crushed to get the gold out of it, and that was expensive. Only after the best of the alluvial ground was pegged did the more experienced men start to explore the reefs.

Three reef miners from New South Wales, Alexander Pollock, his brother Robert, and their mate, Franklin Lawrence, liking the look of some surface indications, started a shallow shaft at the head of the enormously rich Nash's and Sailor's gullies.

The quartz continued from the surface down about twelve feet before slightly changing its line of direction. They sank about two feet alongside of it, and then broke down the rock to see what it contained. Several large chunks of white rock came away and then, with picks still raised, the miners crouched dumbfounded, unable to believe their eyes. The quartz reef they were looking at was a mass of almost pure gold, so solid that the soft, tough metal was holding the chunks of shattered quartz together.

Smashing the rock, and wrenching at the great mass of gleaming yellow gold with their picks to get it free, the three men had soon collected all they could carry. One of them brought their blankets and they collected the scattered lumps of gold and quartz from the bottom of the shaft, wrapped them in the blankets, and quietly carried them away to the calico and bark bank building, not yet opened but already in use, near Acting Commissioner Davidson's humpy.

Word of the find flew round the camp and the diggers crowded in to see, but the manager of the bank dared not put the gold on display because he had no facilities to protect it. The most he would do was to allow a few men into the building at a time to look at it. These lucky ones were besieged as they came out, and accounts of the treasure lost nothing in the telling. On 8 November the *Maryborough Chronicle* reported the finding of a reef "loaded with gold like plums in a Christmas pudding". Miners were quoted as saying that the "Mother of Gold" had been found. The reef soon came to be called the Lady Mary.

The surrounding ground was rushed. On Sunday, 10 November, digger Tom Clarke wrote to his father: "We were on the ground before the Commissioner had pegged off, and could not get a claim."

The Lady Mary Prospector's Claim granted to the Pollocks and Lawrence was 420 feet on the line of the reef and extended right across the heads of Nash's and Sailor's gullies. By 12 November some of the richest stone most of the diggers had ever seen was coming out of it. The line of the reef—running north and south—had been taken up for miles by reef claims each running forty feet on the line and of an indefinite width.

Many of these claims produced rich quartz from the start. Ten ounces of gold were pounded from one block of quartz with a hammer. For most of the distance the surface of the quartz was no more than three or four feet below ground level, the reef was at least ten feet deep, and the gold was within eighteen inches of the top of it. Every specimen was streaked with gold.

A month after the reef was found it was reported that it had been proved rich for nearly 200 yards, and that two miles away gold had been found in this same reef in good quantity.

The Commercial Bank of Sydney was opened for business on 3 December, and on 9 December the manager, Tom Pockley, christened the Lady Mary by breaking a bottle of champagne on it. Formal speeches were made, and a cask of whisky donated by Robert Pollock was broached. Some time after that the whole thing got out of hand, and before the day was done every pub and shanty on the field had made its contribution. Even the New Chums were full of free grog.

The day after the finding of the Lady Mary reef, Frederick Goodchap, Robert Kift and Edwin Morgan, former navvies from the Ipswich-Toowoomba railway, pitched camp on a ridge to the south of the Pollocks, almost on the spot where Nash had sat on his first evening on the field. They were separated from the Pollock's ridge by Sailor's Gully.

After fixing their tents and bunks they began to crack the white surface quartz that was lying about everywhere. They rolled over a big lump of it weighing about a hundredweight, and in the quartz in the ground underneath it saw a streak of pure gold an inch wide. They cracked the quartz with their picks and wrenched out the gold like yellow toffee with splinters of quartz still clinging to it. Before they stopped they had filled a sugar bag.

Gold fever rose to a new pitch in the rush to peg out the reefs with which the whole of the surrounding country was seamed. All over the field everything that looked like a reef was pegged. Some claims were taken by men who had not even the money to buy rations. A visitor was offered a share in a freshly pegged reef for five pounds and refused. That afternoon the owner dug down a few feet and uncovered rich stone. The man who had refused to invest five pounds now offered a hundred, and was refused in turn.

Many of the reefs were fantastically rich. Great lumps of quartz containing nearly as much gold as stone were humped in by sweating miners to be lodged in the bank for safety. There were shafts where one could see gold running through the quartz in broad bands, and others where the owners picked out bucketfuls of specimens in which the quartz seemed to be laced together with a network of gold.

Blacksmith James Cockburn was called to look at a mine after two shots of explosives had been fired. He said it was like a vision. A great band of gold shone across the white face of the quartz like a streak of lightning.

Within hours of news of the Mother of Gold reaching Brisbane all the the hard-luck stories of disappointed New Chums had been swept aside in a mad scramble to get away to the goldfield. There was not a pick or a shovel to be had in town. Men and horses alike disappeared from the streets as every animal that could stand was bought up and loaded with stores for Nashville. Shopkeepers and their assistants, lawyers and their clerks, doctors, parsons and even schoolboys —no matter that they knew nothing at all about reef mining —all packed whatever they thought they might need in the way of rations, clothing, blankets and tools, and took the treacherous track via Durundur Station for the new diggings more than a hundred miles to the north. Every ship in the river was loading cargo for Maryborough. There seemed to be nobody left in town but women and children.

Among those who remained was Catherine Murphy—the future Mrs James Nash—whose invalid father was at that time struggling to make a small store provide a living for a family of ten. Her young brother and some of his mates had run away from school and joined the rush. She later recalled:

> I often amused myself watching from my father's door where I had a splendid view of the road to Gympie, and all the different styles of conveyance—drays and spring carts and horses, packed from head to tail with tin dishes, picks and shovels and blankets and billy cans, all

rattling and noisy, filling the air with their sound as they rumbled along. The men would be talking and laughing as if they were going to a race meeting instead of facing they knew not what privations and difficulties in the way of scrubs and creeks to get through, and almost inaccessible ranges to cross while going by land from Brisbane.

Those who made the trip along the marked tree line to Durundur and then over the ranges into the valley of the Mary River in this second wave of the rush had rough going. One of them wrote to the *Queensland Daily Guardian* describing it as a frightful journey to footmen and horsemen, and to drays almost impossible. He went on:

> All sorts of schemes have to be resorted to to drag the laden drays up the steeps, but lowering them down again is the most difficult. Ropes run around trees have to be attached to the drays, and hand over hand they are lowered, whilst men bear a hand with some guy ropes to prevent the whole toppling down some siding. Notwithstanding all precautions, a good many horses have been lost.

In spite of all the difficulties so many men were on the track that at night time their campfires made a winding trail of light that could be seen leading through the bush for miles. All travelled as light as they could, and most of them carried butcher's knives, for fear, as one of them put it, "of being attacked by ferocious sheep". They rarely went short of mutton, and some squatters started talking about shooting a few diggers. Diggers complained bitterly about lack of hospitality at station homesteads.

On 2 December Mr H. E. King took over as Gold Commissioner at Nashville from Acting Commissioner Davidson, and on 9 December he reported: "With few exceptions, the whole of the miners employed on Nash's, Walker's and White's gullies and the gullies in the scrub are getting gold in payable quantities, while some have been extremely fortunate."

Government Road Overseer Bragg was sent to cut down the trees growing along the path from the Commissioner's camp to Nash's alluvial gold bed and the Pollocks' Lady Mary claim. The branches were burnt, the trunks dragged out of the way with bullock teams, and the stumps left standing. The track was called Mary Street. Before Bragg had finished, far-sighted men were building bark shanties and shops under the shade of the trees which still lined and overhung it.

A growing problem on the field was the safety of the gold that had been mined. The small, slab bank was soon so full of rich quartz specimens lodged there for safe keeping that it could take no more, and all over the diggings fortunes were hidden away in old tin chests, in holes scooped under bunks, at the bottom of shafts and even, it was said, in hollow logs away out in the bush.

The bank, to the disgust of diggers who had been waiting impatiently for its opening, offered only three pounds six shillings an ounce for gold, which was well below ruling prices in the south. Because of this, they sold only enough to be able to buy rations, and continued to hoard the rest until they could take it out themselves. All were pressing for the establishment of a gold escort service, but there was no sign of one being started.

Cases were quoted of men with from three hundred to four hundred ounces of gold leaving for Maryborough at night so that the trip could be made on the quiet, a necessary precaution in view of the goldfield's floating population and rumours of bushrangers in the vicinity. Many who were getting a lot of gold almost stopped work because they could not get their gold away from the field in safety.

The men on the newly opened reef claims were worst off of all, since there was as yet no sign of crushing machinery being brought to the field. The owners of the Lady Mary Prospector's Claim had closed down. Their quartz was so rich that it was not safe to stack it above ground, and the bank had no room to store any more of it. All they could do was protect the claim by occupying it every day, and crushing enough of the gold-laden quartz in an iron dolly to meet their living expenses.

One-time Native Police officer E. B. Kennedy, who was invited down the mine at this period, wrote:

> This shaft was a large, open one; two easy drops and we were at the bottom. On a sheet of bark being removed we were transfixed with astonishment. The slab of quartz exposed was about a yard and a half long and about two yards deep. These were not, of course, its natural boundaries. The quartz was very white, contrasting strongly with the gold scattered all over its surface, chiefly in specks the size of two or three pins' heads, but sometimes in patches as large as a pea. There was scarcely a square inch of quartz without gold in it. On a piece of the quartz being chipped off, we found the gold inside as thick as ever.

More gold-seekers were pouring into the field all the time, and tensions were mounting. The *Maryborough Chronicle* reported:

> There are from 2,000 to 2,500 men on the ground. Some exceedingly rich quartz has been got by some, whilst at the same time numbers of men are said to be starving. There is danger of the stores being rushed and plundered. Many have arrived without a shilling in their pockets and have done no good since their arrival. The great complaint is of the number of people on the field who are not diggers, and who are without the ways and means of living.

On 12 November a private gold escort of four men left the diggings for Maryborough with five hundred ounces of gold. Few saw it as a solution of their difficulties, and a newspaper correspondent commented wryly: "The Maryborough volunteer escort, as at present composed, has not the prestige of invincibility about it to lower our apprehensions, and I would not promise them therefore, a complete loading."

The same issue of the newspaper carried a report from further north that was even less likely to inspire confidence. It was of the murder and robbery of two gold escort troopers on the track from Rockhampton to Clermont. As the story behind the robbery was revealed the whole colony was shocked by its amazing brutality and cold-blooded planning. To many of the Nashville diggers it had a special significance, as it concerned a man they had known and detested.

Chapter Eight:
The Escort Murder

In 1863 there arrived in Clermont as Gold Commissioner and Police Magistrate a man many soon felt they would have been better without. This was Thomas John Griffin, aged about thirty, a handsome, blond-bearded giant of a man, accomplished in the military arts, claimant of aristocratic connections, and a great gambler.

He was said to have been born in Sligo, Ireland, son of an English army officer, to have joined the Irish constabulary, and served as an officer in the Crimean War before coming to Victoria as a member of the colony's police force in 1856.

In Melbourne he married a widow older and wealthier than himself, left her when the money ran short, and after a supposed sojourn in New Zealand during which his wife was shown a newspaper notice of his death, he arrived in Sydney, presented his credentials to the Police Department and, towards the end of 1858, was sent to Rockhampton as Chief Constable. After the separation of Queensland from New South Wales in 1859 he became Chief Constable in Brisbane, and then Clerk of Petty Sessions. He was believed, at the time, to be courting the sister of a member of the Queensland Government. His promotion was rapid, and in 1863 he received the appointment to Clermont, then a thriving gold town.

To his superiors, of whom there were few in Clermont, Griffin always behaved with great charm and polish; to his inferiors, of whom by the standards of the time Clermont possessed many, he was intolerant and arrogant. In one matter he made an exception: when short of money, which was often, he would borrow from anybody foolhardy enough to oblige. He spent a large part of his time in a well-known gambling establishment.

Before long Griffin was deeply in debt, and heartily detested by most of the diggers who not only disliked his manner but claimed he allowed his indebtedness to influence his decisions on the Bench.

In September 1867, at a public meeting presided over by the mayor, a resolution was carried, "that it is desirable for the welfare of the community that Thomas John Griffin should be removed from his position as Police Magistrate in consequence of the inefficient and unsatisfactory discharge of his magisterial duties". Signatures were collected, a petition presented to the Government, and Griffin was transferred, as Assistant Gold Commissioner, to Rockhampton, where Commissioner John Jardine could keep an eye on him.

A select group of his friends gave Griffin a farewell party, but sorriest of all to see him go were six Chinese who, several months earlier, had given him some money and a parcel of gold to send to Rockhampton by the gold escort, and had heard no more of it.

Griffin arrived in Rockhampton from Clermont on 19 October 1867, two days after Sergeant Julian, of the gold escort, had ridden in from Clermont with another parcel of 2,806 ounces of gold. There Griffin received news that six Chinese diggers from Clermont were in town waiting to take ship for China, and they claimed they had been told that money and gold they had handed to Griffin at Clermont had never reached Rockhampton. Griffin found himself in need of money in a hurry.

The money obtained from the sale of the 2,806 ounces of gold brought down by Julian was due to be escorted back to Clermont. Griffin, though he no longer had any authority to do so, instructed Sergeant Julian to escort the money back and said he would come part of the way with the escort.

Julian quickly became suspicious of Griffin, but his position was difficut. Escort troopers were a special force who came under the direct control of the Gold Commissioner and, Commissioner Jardine being temporarily absent, Julian came under Griffin's authority.

From that point on the escort to Clermont became a story of one clumsy attempt after another by Griffin to get the money in such a way that it would leave suspicion on somebody other than himself. Sergeant Julian, and Trooper Cahill who accompanied him, outwitted the former Commissioner at nearly every turn, but they took no other action. Griffin was known to have influential connections, and they no doubt were thinking of their own careers.

After a good deal of shuffling about, apparently caused by the escort men's distaste for the job, the escort, consisting of Sergeant Julian, Trooper Cahill and Griffin, set out from the Rockhampton escort camp about 4 p.m. on 27 October, Julian carrying the money, totalling £8,151, in a parcel of ten canvas bags.

The Chinese diggers, meanwhile, had made several approaches to Griffin for their money, and he had put them off with promises of payment.

The first night out the escort camped at Stanwell, about eighteen miles to the west of Rockhampton. Julian scratched a hole in the ground, put the money in it, laid his blankets over it, and slept on it. About 3 o'clock in the morning Cahill went to bring in the horses. Julian noticed that Griffin seemed to be watching him carefully, so he took out his revolver and pretended to be cleaning it. Cahill returned with the horses soon after.

As they prepared to break camp Griffin seemed reluctant to move. He asked Julian how much money he was carrying, and when told "over eight thousand pounds", said it was too much for a small party to take through dangerous country and they should go back. Julian disagreed. Griffin said the horses needed to be shod, and that he and Cahill would take them back while Julian remained in camp with the money. Julian refused to be left alone with the money, and finally all three rode back to the escort camp, which was only a few hundred yards from Rockleigh, the residence of a Mrs Ottley, whose daughter Griffin was currently courting.

At the escort camp they found Troopers Power and Gildea. Cahill and Power were sent to town with the horses, and after they were well on their way Griffin told Gildea to go to Rockhampton for the mail, and himself left for Rockleigh to see the Ottleys without telling Julian what he had done. Julian saw Gildea getting ready to leave, asked where he was going and on being told, instructed him to remain in camp until he brought Griffin back.

Griffin, fuming with rage, returned to the camp with Julian and Gildea departed for the mail. Terse words were exchanged by Griffin and Julian as they waited, but not until the troopers had returned from town would Julian agree to Griffin's going back to the Ottleys.

Griffin returned to the camp in the early hours of the morning but went back to the Ottleys again before breakfast. The reason for the visit became fairly obvious when Julian, suspicious about the look of the tea as he tossed it into the boiling billy, took a sip of it and found it tasted bitter. He tipped it out and saw traces of white powder in the bottom of the billy. When the troopers came up with the horses he told them he had spilt it by accident, that there was no time to make more, but there was plenty of milk to drink anyway.

After breakfast Griffin joined them and led the escort off on what he called a short cut through deserted swamps and scrubby country towards Gracemere Station. Julian noticed him looking back all the time as though expecting something to happen. If the men had drunk poisoned tea and died in this kind of country, their bodies could easily have been hidden so they would never have been found.

Getting near Gracemere, and all the party apparently quite healthy, Griffin said he had left behind a small parcel of gold that had to go back to Clermont. He sent Cahill back to the escort camp, and after riding a little further, told Julian to go back to the camp and wait with the others.

Instead of waiting at the camp, Julian collected the others, rode into town with them, and put the money back in the bank. Griffin arrived in time to see them coming away from the bank, and furiously demanded to know what they were doing there. Julian said he had deposited the money in the bank for safe keeping. Griffin bawled him out, suspended him, and appointed Trooper Power to head the escort in his place. He then took Power to the bank, told the manager, T. S. Hall, what he had done, and asked that the money be given to Power. Hall, in view of Power's inexperience, gave him half the money only, and Griffin dispatched Power back to the escort camp with it. Then, unknown to the others, he met the Chinese and told them he would pay them the following day.

After that, Griffin rode out to the camp, told Power he would take care of the money that night so Power could get a good night's sleep, and took the parcel with him to the Ottleys.

Next day, while the troopers remained waiting at the camp, Griffin went to Rockhampton and paid the Chinese the money owing to them—a total of £252. The following morning, 31 October, he returned to the camp and handed the money packet to Power who, when putting it in his saddle bag, felt a vacant space in it as though one bundle of notes had been removed.

Trying to gain time, the trooper told Griffin one of the horses had a sore back and was not fit to travel. Griffin ignored him and told Cahill to go and round up the other horses. Power called out to Cahill in Gaelic, which both of them but not Griffin understood, not to bring the horses but to drive them away into the bush.

Cahill disappeared and returned at last to say he could not find the horses. Griffin, furious, ordered him back to look again. Power pressed the matter of the horse with the sore back and was told to take the animal to town and get something done about it. Griffin took the money packet in the meantime.

As soon as Power got to town he went to the bank and asked Mr Hall to come out to the camp and see that the money was all right. Hall came next morning and suggested sealing up the mouths of the bags in the money packet, but Griffin opposed the idea, saying it was useless as the seals would break from the friction on horseback. Hall was not prepared to make an issue of it, and left without actually checking the money.

Left to fight it out alone, Power refused to take charge of the packet unless Griffin himself sealed each of the bags it contained. Livid with rage, Griffin had no alternative but to seal the bags, thus making himself responsible—if the bags were ever delivered—for any money that was missing from them.

Then, once more, after delays of nearly a week, the escort moved off—Trooper John Francis Power, who carried about four thousand pounds in notes and coin, Trooper Patrick William Cahill, and their chief, Assistant Gold Commissioner Griffin, who had told Hall he would not be accompanying the escort further than Gogango, about forty-five miles out.

On 5 November, when the escort reached the Mackenzie River crossing, a long way past Gogango, Griffin was still with it. The party made camp about four hundred yards from Bedford's Hotel at the crossing. They dined at the hotel and Griffin told Bedford he was going back to Rockhampton next day. Bedford was going also and the two of them arranged to travel together. The two troopers left for their camp about 8.30 p.m., followed soon after by Griffin who carried a pint bottle of brandy.

During the night Bedford was awakened by a pistol shot about 2 a.m. and again by another about 3.30. About 4 a.m. Griffin arrived at the hotel looking very tired, and said he had lost his way. Asked about the shots, he said he had heard only one, and that was fired by Power when he lost his way when looking for the horses. Griffin and Bedford left the hotel for Rockhampton about 5 a.m.

Griffin insisted on riding behind all the way and was apparently having trouble with his swag, which was large and did not carry well. At one spot he told Bedford to ride on while he went off the road about thirty or forty yards. At that same spot a man later picked up one of the escort's one pound notes. They arrived at Rockhampton on 8 November, and Griffin used another of the notes to pay for some drinks.

That afternoon word reached Rockhampton that Power and Cahill had been found dead, apparently from poison, at the Mackenzie crossing. They were believed to have been poisoned because they had vomited and several pigs had died as a result of having eaten it.

Griffin affected great concern and surprise, but it apparently occurred to him that if the troopers were found to have been poisoned, he would be suspect. Next morning

Hall, the bank manager, remarked to him, "However did those two fellows got the poison; I can't make it out." Griffin forgot himself and exclaimed: "They are not poisoned; it's all a trumped-up yarn—a false report; they are shot and you will see if they are not."

By saying that, he had sealed his fate either way, because if the men were poisoned, he was the prime suspect; if they were shot, how did he know they were shot when everybody else thought they were poisoned?

The police party that left Rockhampton for the scene on the morning of Saturday, 9 November, included Sub-Inspector George Elliott, Detective Kilfeder, Mr H. P. Abbott, who was an inspector of the Australian Joint Stock Bank, Dr David Salmond, Sergeant Julian, and Griffin himself. Griffin, like the police, carried a revolver. The police certainly suspected him, but he was still a dangerous man to make any allegations against.

Throughout the journey Griffin displayed the same bull-headed stupidity he had done throughout. If poison had been given to the two dead men, it was to his advantage to delay medical examination as long as possible. All the party were mounted on horses except Dr Salmond, who could not ride and was given a gig to drive. Griffin offered to have his own horse led and drive the gig for the doctor. Salmond agreed and soon regretted it. Stones, stumps and trees—Griffin seemed to be aiming at them deliberately. At last they came to a creek with steep banks and a log across the track. Griffin, as though in a panic, threw up his hands and in doing so threw the reins up onto the horse's back out of reach. Luckily for Salmond the horse, left to himself, dodged the log and came to a stop on the opposite bank.

Salmond turned to the driver: "Mr Griffin, you are the most careless or the most reckless man I ever saw in my life; I have had many narrow escapes today, but this is the narrowest—and the last."

Griffin looked hurt. "Surely you don't mean it, doctor?"

"I do mean it, Mr Griffin; you will not ride another yard with me."

Griffin remounted his horse. Elliott watched it all, and waited.

On Sunday they stopped at a wayside inn for dinner, and Elliott quietly took the hotel-keeper aside.

"Mr Griffin and I want to have a private room to talk this unfortunate matter over, and we will have something to drink," he said. "I shall always ask for gin, and you be sure to give me water. Give Mr Griffin whatever he asks for."

Elliott's plan worked well. It had been a hot, tiring day, and after several drinks Griffin became drowsy. Elliott lay back in his chair and pretended to doze off, but as soon as he was sure Griffin was asleep he reached across and removed the Assistant Commissioner's revolver from its holster.

The weapon was the old-fashioned type of those days, with separate caps for firing each chamber. Working quickly but quietly, Elliott removed each cap, scraped out all the detonating material, wet both the inside of the caps and the powder in the chambers, replaced the caps and slipped the weapon back in the owner's holster. The job done, he jumped to his feet.

"Come on Mr Griffin, we have been too long asleep; we must get on."

They reached the Mackenzie crossing about 9 a.m. on Monday, 11 November, and put up at Bedford's Hotel. The spot where the troopers had camped was in a little hollow between the road and the river and screened by a patch of scrub.

The bodies, which had been buried, were exhumed, and Dr Salmond began the extremely unpleasant post-mortems alone, with everybody else keeping well up-wind. After he had examined both bodies Sub-Inspector Elliott approached.

"What have you found, doctor?"

"I've found bullet wounds in the heads of both these men."

The policeman kept his face expressionless. "Then don't tell a soul about it for the present."

Elliott crossed to Detective Kilfeder.

"Griffin is the man," he said. "Get into conversation with him and go across and sit on that log. I'll come and join in the conversation. I'll sit on the other side of Griffin, and when I give you the wink, on with the handcuffs."

Kilfeder did his part, and Elliott joined them on the log.

"This is a sickening sight," he said. "Have you got a drop of brandy in your flask, Mr Griffin?"

"Yes," replied his unsuspecting quarry, putting his hand in his pocket.

"Now!" cried Elliott. They grabbed him together and in an instant Griffin found himself handcuffed. To everybody's surprise he took it very quietly.

"Well, I can only expect it," he said. "I was the last person known to be in the company of these poor fellows."

Griffin was tried at Rockhampton and, on circumstantial evidence, found guilty of murder and sentenced to death. The stolen escort money had not been recovered, and he went to the gallows protesting his innocence.

A reward of two hundred pounds was offered for the recovery of the money, and only then was it revealed that Griffin had made one last bid for his freedom.

Two warders from the jail where Griffin had been held pending his execution revealed that he had promised them large sums of money if they would let him escape. He told them he had killed the two troopers, stolen the money and hidden it. He drew maps to show them where it was hidden in a hollow stump. Three nights running the two jailers had gone out looking for the stump, but in the dark could not find it. Disbelieving Griffin's story they said nothing and he was hanged.

The men produced Griffin's maps and, in broad daylight this time, led a search party to the spot Griffin had described. They found the stump and inside it the missing escort money. They were paid the two hundred pounds reward for recovering it—and sacked for not reporting what they knew sooner.

Then came the grim story's most macabre twist of all.

Because of all the bizarre circumstances of the case, and the fact that a man in Griffin's position had turned to murder, a number of doctors had shown an interest in obtaining Griffin's head for research purposes. To prevent this, when the body was buried extraordinary precautions were taken against possible grave robbers. A second coffin containing the body of a man who had died on a ship was buried on top of Griffin's, and the grave was guarded night and day for a week. By then it was thought Griffin's body would be safe from interference.

A few days later a rumour was circulating in Rockhampton that Griffin's grave had been opened and the head stolen. So persistent was the story that permission was obtained to open the grave. The body in the top coffin was found undisturbed. Then Griffin's coffin was opened. Griffin's head was gone. Only the headless trunk remained.

The offer of a twenty-pound reward brought no results, though by then it was generally known in Rockhampton what had happened. It was said that a Rockhampton doctor, assisted by a sailor who knew about the second coffin, opened the grave and stole the head. The doctor took it to his house but the smell was so offensive that he arranged for the sailor to take it out of town and bury it in a suburban garden until it could be cleaned. The man collected it and was carrying it through the deserted streets about midnight when he was attacked by a pack of dogs and had to drop the bag and run for his life. There followed a macabre chase in the middle of the night as the sailor and the doctor pursued the dogs through the streets and eventually recovered the head and buried it as originally planned.

Eventually the skull was dug up, cleaned, and given an honoured place among the doctor's curios. Though he often told friends how he came by it, nobody ever claimed the twenty pounds reward.

Chapter Nine:
Shanty Town Christmas

All that December at Nashville there were rushes and rumours of rushes over the surrounding country. On 6 December a party of diggers applied for a prospector's claim on a creek about a mile up-river on the track to Brisbane. Gold Commissioner King went out to mark it off with a mob of hundreds at his heels. "Old men have become boys, and boys wallabies in the jump for fortune," wrote the local correspondent exuberantly.

Every time a man was away in the bush overnight, half a dozen hopeful diggers would be out looking for him in case he had struck gold. Two heavy thunderstorms temporarily relieved the water shortage, and though most of the gullies were too shallow to last, the water in them soon looked like yellow soup from all the dishes of gravel that had been washed.

By 23 December Commissioner King was reporting: "The rush to the One Mile Creek on the Brisbane road has turned out very well. The population in that locality cannot number less than 500 and is increasing daily. The number of real miners is increasing very fast and the population has settled down to heavy work. The gold has been traced up into the banks of the creeks and gullies in the scrub. I estimate the population on the ground at 3,500, and about 500 have left the diggings to go to Maryborough or Brisbane for Christmas."

King's estimate of the population was probably a good deal short of the actual figure. Enough cattle were being slaughtered to provide beef for twice as many, and there were supposed to be about fifty wagons drawn by a total of about 250 bullocks and horses constantly on the road from Maryborough with supplies. The country around the diggings was full of prospectors nobody in the township ever saw. A man would come in with a pack-horse or two, buy rations for half a dozen men or more, and disappear into the bush again with nobody knowing where, or how far, he was going.

Nash claimed there were between fifteen thousand and sixteen thousand people in Nashville that first Christmas, and allowing for those who came down from Maryborough and flocked in from the surrounding countryside, his estimate was probably closer to the correct figure than that made by King a few days before. However many there may have been, it seems certain they all had a fairly lively time. The *Queenslander* correspondent, as usual, was in the thick of it. He wrote:

> Christmas eve at Nashville was given up to perambulating Mary Street from one end to the other. The Milky Way was ankle deep in dust, but it mattered not to the tide of life that rolled over it. Shops and stores were crowded, some even into the street.
>
> Of the public houses, some gave forth their vocal strains, others attracted with performances of the light fantastic toe, whilst others again staked their chances of success upon the attractions and persuasive powers of comely maids told off to preside over the decanters.
>
> Butcher's meat hung in profusion at every dozen paces on either side of the street. Being half bled and green crested, the beef, at most, found sale as on ordinary occasions and, there being no attempt at lighting up beyond the dim flicker of the slush lamp, nothing complementary of the meat stores transpired.
>
> The twenty-fifth broke upon us with as many choruses as years in the present era. Hundreds who had lain about in the toilets on the previous day opened their eyes to the morning sun in partial consciousness, and drowned the spectre of their imagination in a fresh libation.
>
> The day thus begun ended in many strange freaks, revelries and excesses which, being shared in by hundreds and witnessed by thousands, made Nashville a

place to be remembered. Sections of the more circumspect held religious services.

Boxing day was celebrated with foot racing, jumping, etc., but the largest business was done at the fountains of spirituous enjoyment. Every hour wound up with a free fight in which no one was hurt. Three balls and Old Tom suppers were provided for the evening.

On the 27th all was quiet and an enormous stillness succeeded the excitement of the two former days. Business men complained at the dullness in trade, and only in sticking plaster, pills and quinine were there signs of activity.

It was not until a day or two later that stockmen from surrounding stations mustered the strength to mount their horses, and visitors from Maryborough turned up to catch the early morning coach home.

The track carriers cut through the scrub and the crossings they laid over creeks brought the travelling time between Nashville and Maryborough down to a couple of days for a light vehicle. H. La Barte ran a four-horse coach service, leaving Nashville at dawn and arriving at Maryborough—weather permitting—about dusk the following day.

Nashville, by this time, had become a large, ragged-looking settlement, swarming with people all day and most of the night. In all directions lines of tents and bark huts stretched out into the scrub, winding along to follow the gullies and ridges, clustering on promising flats, and dotted among the tall trees that still covered a large part of the diggings.

The country was all hills and hollows, and in later years it was to be said that, like ancient Rome, Gympie was built on seven hills—Surface Hill, O'Connell Hill, Commissioner's Hill, Calton Hill, Caledonian Hill, Palatine Hill and Red Hill.

Diggers' tents ranged from a tarpaulin thrown over the low branch of a convenient tree to elaborate canvas houses. The huts were mainly of bark with thatch roofs, though a few were more substantial with split slab walls and shingles.

Much of the early Nashville building was done by the Aborigines of the area, who had orginally built fairly substantial bark and thatch huts of their own. To get material for these the natives chopped rings round the ironbark trees that made a large proportion of the forest, then cut a line down the side and peeled the bark off in a single sheet measuring about six feet by four. Pressed flat and left to dry, the bark became as hard and durable as timber—strong enough, when laced or nailed onto a sapling frame, to keep out the wind and rain for years. The price of one of these Aboriginal-built homes was generally about three shillings, payable in tobacco, sugar or rum.

The most substantial building on the field was an old log cabin, obviously dating back to the earliest days of white settlement, which somebody had discovered among the trees. Far more strongly constructed than any ordinary shepherd's hut, it had walls of solid logs notched together at the corners, and nicked to make loopholes for gunfire. The heavy door had a strong bar to bolt it firm, and the ironbark roof was strong enough to have stopped a hail of spears. The interior was divided into two large rooms, both earthen floored, and one room had a large fireplace across the end of it.

So well built was this log cabin that it survived into the 1890s, by which time most of the early goldrush buildings had disappeared. How its unknown owner had lived in the locality and never picked up any of the gold that lay all around him remained as great a mystery as his own identity.

Mary Street, two months after the rush began, was about a mile long and was supposed to be forty feet wide, but there were still a lot of tree stumps in it, and the motley of bark and slab buildings that lined it were built as and where the owners had felt inclined. Some shops jutted half-way across the roadway, others sheltered back among the trees. Signs painted on calico were strung among the branches and notices on bark nailed to the trunks. Building sites were already changing hands for as much as sixty or eighty pounds.

In places the diggings encroached almost onto the roadway, and at the lower end, where Nash's Gully ran across it, the shafts—some of them sixteen to twenty feet deep—were so close to the dray tracks that there was only just enough room for one dray to get through.

The *Queenslander* correspondent described a walk up Mary Street from the ridge on the Maryborough side of the township.

> Towards the left you find a regular line of shanties running parallel to Nash's Gully. Descending to the flat, you pass through groups of men, some at work singing, others at the windlass, and strollers who have stopped to examine and give an opinion on the wash dirt hoisted up.
>
> You pass Mr Gilbert's saleyards on the left, very substantially erected, and just above on the rise, the Commissioner's tent, used as an office and over which waves the Union Jack. Keeping to the left of the street you pass a whole lot of stores—Mr Tom's canvas establishment, Croaker's spirit store, J. and J. Burns', Booth's Post Office and Government Savings Bank stores, and Bytheway's. Rose's restaurant comes next, at which the hungry visitor or digger can refresh himself with a hot meat pie or roast beef, potatoes and duff, all good, and cleanly served for at most only one shilling and sixpence.
>
> Mr Southerden's slab store is one of the best on the diggings. It is entirely built of broad, sawn slabs and roofed with shingles. On again, we pass Rice's, Markwell's, and Graham and Co.—a good store, the sides of which are of galvanized iron—then Ahern's, May's grog shanty, and Hulyer the butcher. At the back of the butcher shop, on the rise, stands a baker's oven and shop, and here a loaf is to be had for sixpence. A few yards from Hulyer's stands Brown's forge, kept going all day from picks requiring pointing and horses wanting shoes.
>
> Now we proceed down the other side of the street nearest the creek. Opposite Ahern's is a pawnbroker's shop, next to which stands Wilson and Co's. tent. Next comes West's public house, a large and strongly built place containing two sitting and four bed rooms, the bar being large and roomy. Mrs Goodwin has a bark place between West's and another place owned by a person named Hyman, who is the cheapjack and funny man of the place. As soon as you come under his eye he has a slap at you, and should an unlucky digger try to take a rise out of him, he is bound to get the worst of

> it, much to the amusement of the crowd which is soon collected. Mason has opened a pub adjoining Hyman's and has no doubt had an eye to business in so doing, for after a crowd disperses who have laughed heartily, many require to wet their whistles and so turn in to the first place of refreshment.
>
> Other smaller shanties are erected all along the creek, interspersed with miners' tents and the shades which are put up over the holes in course of being sunk.

All through the day the ridges and gullies were like a gigantic ant-hill, with men winding buckets of dirt out of hundreds of gravel-circled pits by windlass, bagging it, and carrying it down to wash by the river. A man could not hear his own voice above the scraping of hundreds of shovels and the constant "click-clack" of the cradles.

The alluvial workings had expanded so rapidly that there was a shortage of both cradles and timber, and old packing cases were being bought eagerly by men making their own. A former carpenter, labouring to make a living from a poor claim, walked off it, returned to Maryborough for his tools and a load of sawn timber, and before long was making better money building cradles than many of his customers were making by digging gold.

Work on the field generally finished for the week on Saturday afternoon, and on Saturday night the diggers relaxed. Though a few other wives and families had followed Mrs Leishman to the field, women were outnumbered by hundreds to one, and they were eagerly competed for and fought over.

Most hotels had dance halls of some sort attached to them. There were piano and fiddle bands and flocks of attractive young dancers and barmaids to entertain the diggers, perform rollicking can-cans, and generally encourage the customers to part with their gold. Before the evening was far advanced golden nuggets would be rolling on the tables as the lucky ones competed for their favours.

In the Appollonian Vale Hotel Billy Barlow ran a free-and-easy with a stage, and tables placed all over the hall. Entertainment ranged from girls to wrestling. Billy himself had a flair for composing ribald verse about local men and events, and he also sang from a considerable repetoire, favourites being such hits of the day as "The Blue-tailed Fly" and "Granny Snow". Admission was free, but the liquor expensive.

The proprietor of one of the shanties had as his star attraction a seductively buxom young charmer called Goldie who, though she was regarded as his wife, still managed to keep the place crowded with diggers seeking her favours. Both of them prospered exceedingly until one day a stern-faced woman got off the coach, made some inquiries, and soon after came striding in the door of the shanty.

She announced herself as the proprietor's wife, demanded to know who "that woman" was and, getting a stunned stare instead of an answer, screamed, "Get that Jezebel out of here", and advanced on Goldie with murder in her eyes.

Several diggers sprang to restrain her. Others cried, "Let 'em fight it out."

Diggers took sides, and in the ensuing fracas the lamp was broken, the shanty set on fire, the liquor was drunk, and the proprietor and many of the participants carried down to the river and tossed in.

By the following night the proprietor and his wife had both left town—in opposite directions, it was said—but Gympie's Goldie stayed on.

No work was done on the diggings on Sunday. It was a day to tidy up and wash the clay-soiled flannel shirt and moleskins, to have a look round at how other diggers were doing, to down a few noggins and watch a few fights.

The first church services on the diggings were conducted under shady trees, and one of the early preachers was said to have been a Church of England clergyman who was found working a claim near the One Mile and persuaded to take up his old job on Sundays. Congregations were small and their attention was chancy. The *Queenslander* correspondent reported:

> On Sunday last there was a most extraordinary scene in Nashville. A clergyman was preaching to a congregation of some hundreds who, to all appearance, were most orderly and devout. But suddenly a rumour reached their ears telling them of a glorious fight some short

distance off, when away went one, then another, until nearly two-thirds of the assembly had gone to see fair play. The fight lasted some time in full view of the reverend gentleman until the combatants had found that both were best and given it up as a bad job.

Sunday was, in fact, the traditional time for settling disputes of the previous evening. An old digger recalled:

> It was on the One Mile, Gympie, one Sunday morning. I met a crowd of six or seven hundred people walking up the gully and joined it. As the foremost group arrived at a level spot, one of them looked at it admiringly and said, "We can have a nice comfortable fight here!"
>
> This was the first time I had heard of a fight being comfortable, and I asked a man who the speaker was.
>
> "Oh, that's Doctor Hamilton," was the reply. "He's going to fight Bill Scott, the West Coast champion. Billy's crowd mobbed a man at Huey's last night. Hamilton jumped in and took the man's part, and then they mobbed him. So he is singled out as king of the miners' crowd. But I'm afraid Doctor Hamilton is not in it; he got fearfully kicked by the mob last night, and one of his arms is disabled."
>
> A ring was formed. The men stripped themselves to their waists, the seconds stepped in, and the fight commenced. Hamilton from the start was seen to have the use of only one arm, and many onlookers—who were perched up trees and on every natural vantage place—shook their heads at the idea of his tackling a professional pugilist at such disadvantage.
>
> One hour and twenty minutes of hard fighting ensued, at the end of which time Scott lay unconscious on the ground, badly punished. Scott was carried home. Hamilton walked behind him, dressed his antagonist's wounds and cured him.

Doctor Jack Hamilton eventually became a legendary figure on half the goldfields of the colony. He was not a registered doctor, but had watched his father, who was. When doctors gave him up as hopeless after he was injured in a fall down his shaft at Crocodile Creek he patched himself up and recovered. After that he treated injured diggers wherever he went. He arrived at Nashville in the first days of the rush as a young man in his middle twenties, and was soon known affectionately as "Doctor Jack", always an easy touch if a man was down on his luck, but a dangerous man to run foul of. He was as good at wrestling, swimming, shooting and fencing as he was at boxing, and he had a general knowledge that seemed to cover everything.

He was later to be persuaded to come back from the Palmer River to contest the Gympie seat in the 1878 elections. He won, and represented the diggings for the following five years.

An early Nashville mate of Jack Hamilton's who himself became famous in later years was James Venture Mulligan, already a veteran prospector, though not a lucky one when it came to getting onto a good claim. He had come to Australia from his native County Down, Northern Ireland, in 1859 and followed the diggings in Victoria and New South Wales before coming to Nashville. Doing no better there than he had done in the south, Mulligan eventually rolled his swag and headed north, first to the Etheridge goldfield, at the base of Cape York Peninsula, and then further north again into unknown country. The next thing his old mates at Nashville heard of him was that he had found gold on the Palmer River, in the heart of unknown jungle country, and that there was a big rush on. Scores of Nashville diggers joined it. But that was still a few years in the future.

Chapter Ten:
The Big Nugget

The year 1868 began well for the diggings, with the old ground still yielding prolifically and new finds being reported almost daily.

On the One Mile flat there was only a couple of feet of ground left standing between the claims to mark the boundaries, and at night, according to one digger, "the scene was like a bivouac on the field of battle; all through the night gun and pistol shots were heard, giving notice to marauders that the miners kept their powder dry."

The main diggings were on several ridges which sloped down to the Mary River where it made a bend to form a kind of pocket. Several ridges met there, and between them were gullies and creeks, more or less steep, and all nearly dry. Many claims were worked out and deserted already, their open shafts looking from a distance like freshly dug graves. Between them were waterholes with water yellow and thick as pea soup from all the dirt that had been washed in them. There were tents everywhere. Here and there, incongruously, small parties of naked Aborigines could be seen with an old tin dish or even a piece of bark, busily washing the refuse of some digger's dirt heap, curious to see what it was the white men had been seeking.

Shafts being worked generally had a temporary roof of bark over them to shelter the man winding up the buckets of dirt by windlass and tipping them out onto a stockpile ready to be taken away and washed.

Owners of one-horse drays were making from three to five pounds an hour carting the dirt to the Mary River where, as far as the eye could see, was a line of men, puddling the dirt and rocking the cradles. Each dray was backed to the river brink and its load tipped down a channel cut in the bank to make a heap at the feet of the washers.

The procedure was first to shovel it into long, hollow tree trunks filled with water. There it was raked about, or "puddled" to break up lumps of clay which might contain gold. Next it was washed again in the cradles to get rid of the gravel, and in the process any nuggets were picked out by hand and dropped in a billy at the washer's feet. The fine concentrate from the cradles was then washed by dish at the water's edge to separate the fine gold. One visitor counted more than thirty drays lined up at the river bank at one time.

Some diggers were stacking their dirt at the top of their shafts until they could get it to water. Normally no gold could be seen in these heaps, but after a shower of rain one could see the gold in them glistening in the sun.

There were a few places on lower ground where men had brought water down from running creeks in wooden troughs and were doing good business bringing wash dirt to it in drays and puddling it for the diggers at seven shillings and sixpence a load.

A visitor described an alluvial shaft: "It was four feet by two, and twenty-four feet deep. I was permittted to crawl into the drive and pick out some dirt; and having the luck to find a small nugget weighing about five pennyweight, was allowed to retain it. This man was clearing about thirty or forty pounds a week, but from the very short acquaintance I had with the work, I am satisfied alluvial digging is tremendous fatigue—at the bottom of a deep shaft, lying in a hole which just fits one, and picking the earth away within a few inches of one's face."

In the bottom of the gullies some of the digging was very tough. A man could swing his pick hard and steadily all day in a shallow trench and have hardly any progress to show at the end of it. Diggers came up from the shafts at the end of the day covered in yellow dust from head to foot.

But in the good ground yields were high. An ounce of

gold to the dish was not uncommon, and many substantial nuggets came to light.

With rumours of rushes still rife someone cashed in on the mood of the moment, and the *Nashville Times* reported:

> A heartless hoax was perpetrated by a scoundrel here. On Sunday afternoon the fellow waited on the Commissioner with many entreaties for secrecy and confided to him that a splendid goldfield lay about twelve miles from Nashville near the Maryborough road. He had obtained there, he said, an eight pound nugget and he applied for a prospector's claim. He appointed to meet the Commissioner on Monday and convey him to the spot.
>
> It would appear that he had scattered the intelligence through the town, for many hundreds of men, armed with pick and shovel, were seen pouring along the Maryborough road at an early hour on Monday morning.
>
> The man failed to keep his appointment, so the Commissioner, after waiting a couple of hours, rode out attended by an enormous cavalcade of anxious and expectant men. When they had ridden about nine miles they found a crowd of hundreds waiting. The Commissioner explained the circumstances. Nothing could exceed the intensity of anger displayed by the surrounding mass against the cheating ruffian. Had he been found I have little doubt of the time and manner of his exit from the scene of his audacious exploit. He had obtained a quantity of stores on the strength of his "discovery".

The richest workings were in Nash's, Scrubby, Sailor's, White's, Walker's and Nuggety gullies. The high ground was nearly always the best, yields dropping off closer to the Mary River. Nash and his mates were doing well. By April he, his brother John, Billy Malcolm and Leishman were reputed to have taken an enormous quantity of gold out of their ground. They now washed their dirt at the Mary River, and every week brought the gold to the bank in large leather bags. It was said they kept as much hidden as they brought in.

Walker's Gully had received its name from a character known as Alligator Walker who, prior to the rush, had been shooting crocodiles on the Fitzroy River on such a scale that there were said to be hardly any left there.

He and three mates arrived on the diggings early and staked a very good claim on the gully which yielded heavily. Alligator had been in the habit of drinking the proceeds of his sales of crocodile skins as soon as he received them, and he saw no reason to change his habits when he found himself with his pockets literally bulging with gold. The result was that after the first few days he was rarely sober and more of a hindrance on the claim than a help. There were quarrels, and at the height of one of them Alligator told his mates he was not going to spend the rest of his life grubbing in the ground, and walked off.

They were glad to buy him out, and Alligator used the money to buy a small cutter which he fitted out with the best of everything and named the *James Nash*. He made a comfortable living for a few years trading and fishing before eventually taking up land in the Burrum district.

Parts of Walker's Gully were fantastically rich. The story was told of two young Englishmen who worked a claim on the gully for about four months before walking off it and taking ship from Maryborough for Sydney on their way back to England. Before leaving they confided to a friend that they were taking with them about a hundred pounds weight of gold each which they had kept buried under their tent while giving surrounding diggers the impression that they were getting barely enough to buy rations.

On Saturday, 18 January 1868, there came a report that a man named Murdoch and his party had picked up about eighteen ounces of gold from an excavation on a flat in Deep Creek, not more than six hundred yards from where it joined the Mary River just beyond the One Mile and nearly a mile and a half upstream from Nash's Gully.

That same day heavy rain set in and all the following Sunday it poured. Digging came to a standstill, but nothing could stop the rush to Deep Creek. From all over the field they came squelching in by the hundred to flounder and slide down its steep banks and push their pegs into the mud.

With rain still falling in torrents on Monday, owners of one of the newly pegged claims left it for a few hours, and four other men promptly jumped it. The original owners complained to Commissioner King who held a court in the biggest tent handy, found against the jumpers and ordered them to get out. They refused. Next day King arrived with three troopers, arrested them, and sent them off on their way to Maryborough under escort for trial. The waterlogged diggers cheered their going. If there was one thing they particularly admired in a Gold Commissioner it was prompt action.

The stoppage caused by the weather, combined with the promise of the new field, brought good business to the pubs and shanties. On Tuesday morning a digger was found dead in the water in the bottom of a shaft not far from the Prince of Wales Hotel. Evidence was that he had been seen to stagger out of the hotel the previous evening, very drunk.

All the creeks were soon running bankers and the Mary River itself began to rise. Most of the diggings were within a few hundred yards of it. Many of the shafts were flooded, and most of the others were in danger of it. But though a large part of the diggings was silted up by the floods, few were losers in the long run.

With plenty of water at last, diggers were able to scoop up soil all along the banks of the formerly dry creeks, shovel it straight into their cradles and wash it on the spot. Many quick fortunes were made, but on the other hand, a lot of fine gold went straight through the cradles of the impatient ones to be carried by the rush of water down to the bed of the Mary River. A tremendous amount of Nashville's surface gold was either won, or lost beyond recovery during that first wet spell. Years later the bed of the Mary was dredged for gold lost in the tailings of the rush era, but only a very small proportion of it was ever recovered.

A peculiarity of the alluvial ground at Nashville was its shallowness. In some of the gullies—like Nash's—the wash dirt actually lay on the surface, and many found that the deeper they dug, the less gold they got. There were even stories told of fowls killed for eating having their crops full of small pellets of gold.

Four or five feet was the average depth around Nash's original spot, and even that was when the works were carried into the side of the hills. Big lumps of gold were picked out of little pockets in the rock with nothing more than the digger's bare hands or a sheath knife. Old diggers described nuggets lying about like pebbles or small potatoes, only barely covered with dirt or gravel. The largest nugget of them all was no more than a few inches beneath the surface.

The find was made on 6 February 1868 by George Curtis, Scab Inspector, of Maryborough, who had decided to try his luck on the diggings during his leave, and on 2 December 1867 entered into partnership with Michael Canny to work an old claim at Sailor's Gully owned by Charles Collin. Collin was to receive a third of the proceeds free of all charges and Curtis and Canny the rest.

They began digging on 3 December, but Canny, according to Curtis, worked for only a day and a half before getting fed up, declaring there was no gold in the claim and going home to Maryborough. Curtis wrote to his nephew, Valentine Brigg, who gave up a job on Peak Downs and joined him on the claim on 4 January. Others had worked it with little luck and for the first few weeks it looked as though they were going to do no better. Up until 6 February they had found hardly any gold at all.

Stories vary as to exactly what happened then, but apparently Curtis threw out a bucket of water at the end of the day, and then paused to take a second look at a small section of rock from which the water had washed the soil. He took a tentative hit at it with his pick. The point sank in and stuck firm. Curtis tugged to free it, looked up, and yelled for Brigg.

"Hey Val, come here quick. I've got a nugget, a huge nugget, the biggest you ever saw!"

As Curtis scraped the soil away and diggers clustered round to see, estimates of the nugget's size mounted moment by moment. "Seventy pounds if it's an ounce." "More like seventy-five—eighty, at least."

At last the great lump of gold was freed from the ground. Curtis put it in a bag and humped it on his shoulder to the Commercial Bank.

Next morning the abandoned ground all around the claim was swarming with diggers who pegged out every foot of it and started digging right away. No more worthwhile gold was found there.

Curtis and Brigg worked hard on the claim for a few days without further result and then, on 12 February, Curtis took his find which he had called the Perseverance Nugget to Maryborough where it was exhibited at a charge of a shilling a head, proceeds going towards the building of a hospital at Gympie. It was later shown at Brisbane, and then the Duke of Edinburgh—brother of the future King Edward VII—who was visiting Sydney at the time, expressed a wish to see it. So the big nugget, which by then was generally known as the Curtis Nugget, was taken to Sydney and lodged at Government House to add further to the worries of those responsible for the Duke's safety. It was later placed on public view in Sydney under constant police guard.

Weights claimed for the nugget in its original form ranged from 1,280 ounces to 975 ounces, but the net yield of gold was 906 ounces and its value at the Sydney mint £3,132/9/9. Many Australian nuggets were larger—Ballarat's Welcome Nugget, found in 1858, weighed 2,195 ounces and sold for £9,325—but the Curtis Nugget was Queensland's biggest.

Curtis himself finished up with only a small part of the proceeds of the nugget. Soon after it was found, he had a solicitor enter into negotiations with Collin, the owner of the claim, as to his share, and Collin received £750 as a result.

Then Canny came looking for a share. Curtis refused and Canny took the case to court. He claimed that he had had to return to Maryborough on business and that Brigg had been employed to take his place. He said he had provided a tent, pick and provisions, and that his equipment had been used, not only in finding the Curtis Nugget, but in obtaining about £200 worth of gold from the claim after that.

Curtis called other diggers to give evidence that Canny, before going back to Maryborough, had told them he was sick and tired of the claim and of Curtis and that he had left both of them for good.

Curtis went on to say that the only equipment and stores Canny had contributed was fifty pounds of flour (which he sold before going back to Maryborough and kept the proceeds), a piece of calico for a tent, some tea and sugar and an old pick which was useless and had never been used. He denied that he had got £200 worth of gold from the claim as well as the nugget. He said the total amount of additional gold was not worth more than fifteen pounds and was not enough to pay for the rations of additional men he had taken on to help work the claim after the nugget had been found.

The legal battle went all the way to the Full Court before finally being decided in favour of Curtis. By then a large part of the proceeds from the sale of the big nugget had gone to pay lawyers.

Chapter Eleven:
Women in the Wet

Within a few weeks of the finding of the Curtis Nugget Nashville had become the liveliest little town north of Sydney. The *Nashville Times*, which began publication on 15 February 1868, reported that 560 business licences and 15,000 miner's rights had been taken out in the four months since the rush started.

Fourteen licensed hotels, many of them run by diggers who had done well on the alluvial, and dozens of shanties were doing roaring business among not only the diggers who were flocking to the field, but station workers coming to town to see the sights and cut out their cheques.

Assured of the future of the place, increasing numbers of successful diggers and business men began building more substantial homes and sending word to their wives and families to join them.

The town the women found had grown up where the good ground was—on the north-eastern bank of the Mary River at a point where its course ran north-west. It already contained substantial slab and paling shops carrying calico signs and fronted by sapling awnings covered with branches to provide shade. The only street was still a wagon trail rutted by wheel tracks, dotted with tree stumps, pitted with boggy holes where stumps had been grubbed out, and backed by apparently impenetrable scrub on both sides.

The heavy rains had turned much of the surrounding country into a quagmire, and the street, churned up by a constant stream of bullock drays bringing in supplies from Maryborough, was even worse. Drays stood bogged up to their axles for days; even cows and horses became bogged. There was a big mudhole at the bottom of the street where La Barte's coach had bogged and where, old-timers solemnly informed newcomers, a loaded bullock dray, team and teamster had sunk out of sight into the mud and had never been seen again.

Men who arrived in town late in the day, rather than take their chances in the shaft-pitted scrub in the half-light, often camped in the roadway. One of them described how he spent his first night in Nashville under a tent fly in the middle of the street with the rain pouring down in buckets and water running down the road and around his feet like a mountain torrent.

Another man described the town as one half mud and the other half mosquitoes and sandflies. There were all kinds of people camped there, he said—"parsons without flocks, lawyers without briefs, doctors without patients, men without wives and glad of it, and women without regrets".

Carrier Charles Barrett arrived with his family overland by bullock dray from the Darling Downs, and when they reached the Widgee crossing of the Mary River, downstream from Nashville, the river was running a banker. Another bullocky had joined them on the trip, and the two of them decided they could get across the river by hitching both teams of fourteen bullocks to each dray in turn. They hoped that by the time the dray was in the deepest part of the crossing most of the leading team would be on firm ground, and able to drag it clear.

All went well with the first dray until this crucial point was reached and the leaders started to climb the opposite bank. The ground was wet and slippery and made worse by the water pouring off the drenched bullocks. As they struggled for a footing in the mud, the second team, anchored only by the floundering bullocks in front of them and the bogged down dray behind, were swept off their feet by the current.

The two teamsters, swimming their horses in midstream, did their best to urge the bullocks on, but it was hopeless. Two by two the bellowing leaders were dragged back into the stream by the weight of the others, until both teams, still yoked and chained together, were a wallowing, terrified mass in the water, rolling over and over each other in the current as they were swept down the river.

Among them swam the bullockies, in constant danger of being impaled by tossing horns or stunned by flailing hooves as they seized every chance that offered to yank out the keys holding the yokes and unhook chains to give the animals a chance to struggle out of the flood.

As things turned out, only one bullock was drowned. The others managed to scramble ashore at one place or another and were rounded up eventually. Most of the loading and harness was lost. Days later, after the river had dropped, the two parties and what was left of their gear plodded into Nashville.

Zachariah Skyring, the man who started the Nanango rush, arrived in Maryborough three days after Nash reported his find, and headed straight for the field. He staked a good claim at White's Gully and as soon as he was well established sent word to Brisbane for his wife, three daughters and ten-year-old son Zachariah to join him on the field.

They came early in January by horse dray via Durundur, taking fourteen days on the road. They would have been longer had they not met up with Walter Ambrose and his family, also travelling by dray. Whenever they came to a steep pinch they harnessed both teams to each dray in turn to pull them up. At night young Zac and the driver, his uncle, camped under the dray, while the women slept in a tent.

Another early arrival among the women on the field was Nash's future mother-in-law, Mrs Murphy, who, not having heard from her runaway schoolboy son, came to see for herself if he was doing all right. She found him on a good claim and doing so well that she herself bought a block of land in the main street, had a paling-walled, shingle-roofed building put up on it, stocked it as a drapery store and sent word for the family to follow.

They travelled from Brisbane to Maryborough by paddle-wheel steamer, and daughter Catherine described the trip from there to Nashville by wagon and horse team.

We started on a bright morning seated on our mattresses, etcetera piled upon the wagon. It was quite comfortable for an hour or so, then we would walk a bit to stretch our limbs and get ahead of the team looking for wild flowers and birds' nests. Towards sundown the driver would unyoke his horses near a creek and make a fire at a big log, boil a large billy of tea and sweeten it with a pannikin of sugar. We then spread a cloth on the grass and had tea. So we went on for three days. The creeks were bad to cross and most of the load had to be taken off and reloaded at the other side.

When we got to Gympie, up on the hill over the one long street, we were amazed at the grotesque looking township of primitive buildings and tents dotted all over the place, and hundreds of men filling in the spaces, making it hum like a huge fair. The place was vibrating with life—music and singing, and men shouting and fighting, as they did from morning till night and from night till morning.

The first night of my arrival I went to the front of the house to arrange a lamp. I noticed a multitude of men blocking the street before our place. I asked a person near me what they were there for. He said, "They've come to see the new girl." I simply fled to the back room. But they were not at all a bad lot. They had no intention of intruding.

Women were such a rarity in the early days of the field that those who were there quickly became used to being stared at. Sometimes a man and his wife taking the air on a Saturday evening would be followed by a crowd of a couple of hundred miners, in no way disorderly or intending to be offensive, but just wanting to look at a woman. Wherever a woman went in the daytime, particularly if she was young and attractive, her approach was likely to be announced by a loud, drawn-out cry of "Joe-e-e-e".

The cry had been brought north by old diggers from the Victorian fields where, in the bad old days, troopers—known as Joes—regularly raided diggers for their licences. At first sight of a trooper, diggers would send up a cry of "Joe-e-e-e!" and those without licences would go for their lives and remain in hiding until the raid was over.

There was no need for this in Nashville in 1868, but the old cry was retained and used instead to signal the approach of a woman so that all could stop work and have a look at her. The cry would follow her all over the field. Most of the women, once they got used to the custom, regarded it as a compliment.

A young digger courting the daughter of the proprietor of the Little Nugget Hotel had a hard time of it. The whole thing became a joke among the diggers, and wherever he and his girl went, their approach would be heralded by cries of "Joe-e-e-e!" There was nowhere within a couple of miles where they could find any privacy.

One of the town's early weddings involved a newly rich digger who had spent a good part of his life in the bush. Thanks to advance preparation by his mates, he got through the church ceremony fairly well, but on the way to the reception he thought better of the whole thing, put down his head, and bolted for the bush.

A team of searchers took three days to run him down and bring him back forcibly to face his bride. But he was never happy to be back, his wife never forgave him for bolting, and the marriage was not a success.

When Mrs Tom Cockburn, aged twenty-two, with her two young children joined her husband in Nashville in May, he met her at the coach and walked her to the Apollonian Vale Hotel for dinner. As they waited to be served, about half a dozen heads came peeping round the door to have a look at the new woman. When those had looked their fill, they were pushed aside to make room for others.

Years later Mrs Cockburn recalled her first Nashville home as a "comfortable little place with paling walls and shingle roof". The floor was bare earth and the windows wooden shutters. A big fire place took up one wall of the kitchen, and furniture was a table made of locally sawn pine slabs, a safe with tin sides perforated with nail holes and two stools. There was an iron four-poster bed, and the miners later made her a crib for her two babies. She used it for years, eight additional children being born to her on the diggings.

The first child to be born on the field was Jane Smyth, on 10 January 1868, but such events were still rare and the next entry was not made in the register until 2 April.

Another Mrs Cockburn did not wait to be sent for. Her blacksmith husband James left Brisbane with his eldest boy about the end of October 1867, saying he would send for her as soon as he was established on the diggings. When no call had come by the end of December she decided to wait no longer and headed north, taking her nine-year-old daughter Margaret and six-and-a-half-year-old son Daniel with her, and leaving five other children behind with friends.

Instead of following the Durundur track, she made up a travelling party with three other women, four children including her own two, and one man, to try an alternative route which was being pioneered by sawmiller William Pettigrew, and cedar contractors James Low and William Grigor. A steamer took them about sixty miles up the coast from Brisbane to the Mooloola River and then another three miles up-stream to Low's Landing which, until then, had been used for bringing out timber.

Waiting there was a bullock wagon to take the ship's cargo of stores the remaining sixty miles overland to the goldfield, and an old spring cart for the passengers. The party and their luggage were crammed aboard the cart, the driver whipped up his horse, and they headed into hilly country, winding among trees, stumps and logs over a track worn by the feet of pack-horses and hopeful diggers. The cart was not meant for the job and after about two miles broke down.

The party, encouraged by the fact that the bullock wagon would be behind them, decided to finish the sixty-mile journey on foot. They left their luggage for the spring cart driver to bring on when his cart was repaired, but Mrs Cockburn insisted on carrying a large carpet bag loaded with cakes, pasties and other delicacies she was sure her husband would have been missing on the goldfield.

Half a day's walking through the bush was enough. They made camp that night to wait for the bullock wagon, which they hoped would give them a lift to the field.

The wagon arrived at last, piled high with stores, iron-rimmed wheels grinding and slipping over the stones, and Jonathon, the driver, sweating and swearing over his labouring bullocks. To their request for a lift he explained that he already had aboard as much as his team could handle, and in any case the top of a loaded bullock wagon in country like this was no place for women and children to ride. At the first rock or stump the wagon hit they would all be thrown off, probably hurt, and possibly killed.

Only then did they learn that this was the first bullock team to come through this way. Jonathon had two Aborigines working ahead of him all the time with axes, cutting trees and scrub back from the edge of the track so the wagon could get through.

The bullocks were slow and Jonathon was helpful in spite of having his hands full, but the women and children were hard put to keep up with the team as they plodded up the rough track, skirting the slopes of steep hills, down into ravines, and through creek crossings where the muddy water came up to their waists so that for a large part of the time half their clothing was sopping wet. The cakes and pasties could not be carried and were soon eaten.

Sometimes the road was so steep that the bullocks stopped in their tracks and, ignoring the lash of Jonathon's long whip and colourful tongue, refused to budge for hours on end. When this happened the women went on ahead, but never too far for the wagon to catch up with them before dark. The scrub was mostly dense, green and beautiful, full of native flowers and birds, but it was also completely unknown. There were Aborigines in this country and nobody knew what to expect from them.

To the children, except towards the end of the day when they became tired, the whole thing was an adventure. They splashed about in the creeks, picked wild flowers to stick in their hair, and looked for kangaroos. But the country they were passing through was not as harmless as it seemed, and once the adventure was nearly fatal.

Nine-year-old Margaret saw a bunch of beautiful dark berries that looked like grapes and pushed among the leaves to reach them. But the "grapes" were Gympie-Gympie berries and the stinging leaves of the tree burned her arms and neck with excruciating pain. Her frantic mother, with the screaming child in her arms, ran stumbling back along the track to the bullock wagon where Jonathon rubbed the stings with a mixture of rum and chewed tobacco which did something to relieve them. But for weeks after, when the sun went down and the air became cooler the throbbing stinging returned to remind the little girl of the dangers of the innocent looking bush.

There were places where the scrub gave way to open timber country. But the going was just as hard. By the end of the first week boots and shoes made for the town were worn out and every step was painful.

About half-way along the road they were caught in a thunderstorm. As it bore down on them, the bullocks stopped in their tracks and refused to move. Jonathon, between fuming at his obstinate animals, told the women to hurry ahead to a place about a mile and a half on where they would find a bark hut.

It was almost sunset, and as the weary women tried to hurry along with the children the storm caught them, pelting down through the tall trees and drenching them to the skin. When they reached the spot where the hut should have been they found that it had been blown down and the bark scattered.

One of the Aborigines from the bullock wagon came running up and, in the pouring rain, collected enough bark to build a rough gunyah to shelter the women and children. Then—nobody ever knew how—the Aboriginal made a fire. The rain was coming down in torrents, the whole forest seemed to be drenched and the man had no matches, but still he produced fire and kept it burning in front of the gunyah all night.

Next morning there was no sign of the bullock wagon. Nobody had any food of any kind. The Aboriginal remained with them, not letting any member of the party out of his sight, and apparently listening all the time for the wagon. It was not until the late afternoon that the man suddenly dropped to his knees and put his ear to the ground. After a moment he lifted his head, took a deep breath and sent an ear-splitting "Coo-ee" along the ground. Then he put his ear to the ground again, listened a moment, and stood up, grinning broadly.

"Wagon coming," he said. "About two hours."

Right on time, the bullocks came in sight, making heavy going on the muddy track. The damper Jonathon made on the Aboriginal's fire was the first food they had had for about thirty hours.

Within four days of Nashville they were overtaken by a horseman on his way up to the field. Mrs Cockburn, abandoning all idea of surprising her husband, sent him a message to come and get them.

When Cockburn received it, the only dray available in Nashville cost seven pounds a day to hire. He loaded it with provisions and set out to meet the bullock wagon. The track was not wide enough even for the dray, and he had to cut his way through as the bullock driver was doing. In due course he collected his footsore family and brought them to Nashville.

Cockburn built a bark humpy for his family. The demand for palings was so great at this time, with so many families coming to the field, that the splitters could not keep up with orders and men who wanted them must—if they had time, which a blacksmith in Nashville did not—split their own. Beds were made of sticks and bark, and mattresses of hessian stuffed with dry grass. A tank was made from one of the large, zinc-lined drapery cases in which goods were then shipped from England. It was surrounded with bark to keep the water cool. Wash-tubs were barrels obtained from the hotels and sawn in two.

Water was brought from the river in two kerosene tins on a goat cart with wheels made of split slabs of wood chopped into shape with a tomohawk. Early arrivals with young children had brought to the field herds of goats which had bred and run wild until the mullock heaps swarmed with them.

Wrapping paper was scarce, and if one went to the butcher's shop for a joint of meat, it had to be carried away uncovered on a skewer. Fresh fish was brought about thirty-seven miles by road from Tewantin, on the Noosa River, by a fisherman named Keyser, at first in a wheelbarrow, but later in a spring cart. There was a baker whose real name nobody knew, but who, for reasons equally mysterious, was always called the Crocodile Baker.

Snakes of every kind, from deadly death adders to large pythons, were always a problem in the bark houses. They came into the unlined, dirt-floored huts and coiled up in dark corners or amid the sapling rafters from which they could be dislodged only by hurling a jug of boiling water over them —itself a hazardous operation if they were high up.

One night a digger living in a tent at Sailor's Gully woke up in the middle of the night feeling a weight on his chest. He grabbed it, felt it was a snake, and hurled it out through the open tent flap.

In the morning a woman who cooked for some of the diggers over an open campfire nearby called him to see what was lying in the still warm ashes of the previous night's fire —a large, well-cooked death adder.

Native cats, in the early days of the rush, stole food from the huts, and also old socks and clothing to line their nests in hollow tree branches, but the marsupials disappeared as the town grew. Their place was taken by introduced rats and mice which, having no natural enemies, soon reached plague proportions.

The first cat was brought to the field by Reilly Duckworth after he and his two brothers struck it rich on the Lady Mary reef. The cat had kittens and the competition for the young mousers was so keen that Reilly kept them shut in his humpy and guarded the door with a loaded gun. He was offered more than their weight in gold for the kittens, but refused to sell. Others, seeing the way the market stood, started bringing back kittens every time they went to Maryborough, and before long the price of kittens was down to seven shillings and sixpence each—about five dollars on modern values.

Chapter Twelve:
The Carnival Year

The diggers' name of Nashville died out in the early months of 1868 as the tents gave way to slab huts, and the town that grew up during the bustling, brawling months that followed was called Gympie, a name that commemorated not the lone digger whose find had brought it into existence, but the deadly stinging tree, already giving ground daily before the advance of settlement.

All that year the feeling of carnival continued as people poured into the town and the goldfield spread. It was a year that began at the peak of the alluvial bonanza, with golden nuggets spilled out on rough kitchen benches, hotel bars and dancers' dressing tables; a year that ended with the steady scrape of gravel in tin dishes as the Chinese washed the last grains of gold out of the well-worked ground.

It was the year of the rich, shallow reefs—the "jeweller's shops"—when a man could begin the day hungry and penniless, and before it was over find himself squatting in a shallow shaft looking at more gold than he could carry away. Gold came suddenly and, as the mines went deeper, death came suddenly. A noisy procession coming up the street might be bringing a newly won fortune to the bank; a quiet one, the body of a miner crushed in a cave-in.

It was a year of frantic activity, of rumours and rushaways into the surrounding hills; a year of robberies, bushranging and murder.

It was the year when the old-timers came flooding up from Victoria; when families not used to mixing with all kinds of people were thrown together; when women clashed; children fought; wounds were healed and a new community was evolved to populate the goldfield. The growing town seemed to be bursting with vitality.

Saturday was still the night when everybody came to town, but the range of entertainment had broadened, and soon after sunset town criers with bells began parading Mary Street announcing what was to be seen. The original crier was Walker, who had practised his profession in Brisbane. With large belly, fruity voice and dignified tread, he announced his wares in the traditional Old English style.

"Oyez, oyez, oyez! Ashton's Royal Anglo-Saxon Circus, for one night only prior to departure for New South Wales. On this occasion will be produced on a scale of splendour and magnificence never before witnessed at the Antipodes. Do not be deluded into any penny café or by any false pretender who advises you to witness any entertainment which would disgrace a common tap room. See the greatest circus on earth!"

Then there was Quigly, a great bellman from the goldfields of Victoria, from Lambing Flat and the Lachlan. He was more restrained than Walker, relying mainly on his clear, articulate and stentorian voice to get his message across.

"Let me draw your special attention to the Varieties Theatre, the greatest combination of talent ever assembled in one theatre for the public amusement, the only real theatre in Gympie. Royalty might sit and be amused at the entertainment given by the Leopolds, for the festering tongue of scandal is never moved even to raise a laugh at the humblest in the land. The entertainment given by the Leopolds is the only entertainment worthy of your notice on the diggings."

With the arrival of wives and families the presentation—for advertising purposes at least—had been polished up a little since those first bawdy months, but Saturday night remained as lively as ever, with bells ringing down the whole length of the street and every house of entertainment full, and roaring with life.

By goldrush standards the diggers at Nashville lived very well. Catherine Murphy described the daily scene.

At sundown the diggers would gather around their campfires and cook their steak. This was mostly done by holding it on a forked stick over the red coals. The air was soon well flavoured with roast bullock and one could almost breathe enough for a meal. Splendid beef it was too, as the stations around supplied the field with cattle at a reasonable price. It cost the consumer about threepence per pound.

All the other necessaries of life were not too dear, so that few, if any, were without good plain food—beef, bread and tea, and soon the Chinese grew plenty of vegetables, for which they never charged a very high price.

After the diggers had satisfied their appetites, they washed and dressed in clean flannel shirt and moleskin pants with a handkerchief around the neck instead of collar, a felt hat—with silk, knitted veil hanging over the face to keep the flies off in the day time and twisted turban style for the evening. This was the ordinary diggers' garb, but the dudes among them wore long sleeved Crimean shirts, tweed pants, silk scarf around the waist, and the usual felt hat and veil. They congregated in the street and in the various places of amusement, hotels, theatres, music halls and dancing rooms which were in full swing every night.

The main street was lit by lanterns hanging from shop fronts and trees, but most who went out at night carried lights made from white glass bottles with the bottoms taken off and a piece of candle thrust into the neck from the inside. Every night one could see hundreds of these flickering lights bobbing about over the flats and among the trees. A man who moved without one ran the risk of falling down an open shaft. There were shafts everywhere, many of them abandoned but left open with nothing of any kind to show where they were.

There were always a few fights going on in town. Crowds gathered and bet on the results. Stray dog fights were bet on, and a domestic quarrel between man and wife always attracted a crowd outside their tent to listen. Thin canvas walls gave little privacy. Crowds in the street took sides and barracked vigorously, called advice and suggested fitting repartee to cutting sallies. Bets were laid on who would win, and when it was over arguments arose as to who had won the bets and more fights followed.

Hotels remained open most of the night. A traveller complained: "The visitor who puts up at one of these public houses need not expect to go to sleep before 3 o'clock in the morning and he must be prepared to be roused up at daylight by the noisy demands of thirsty souls for admittance."

Hotel owners, in turn, complained about the shanty keepers, "While any dirty grog shanty does for an unlicensed house, the police are particularly hard upon the licensed man, expecting from the publicans as rigid adherence to the letter of the Act as though they were in the midst of a long settled community."

Within a couple of weeks of the finding of gold at Deep Creek a separate town sprang up around the One Mile, and French Charley Bouel, formerly of Rockhampton, opened a hotel, store and theatre there with a sign across the front emblazoned with the old motto, "Live and Let Live".

French Charley's theatre ran mainly to heavy drama with prima donna Miss Gardiner appearing nightly, but the rest of the etablishment catered for every possible goldfield requirement. Charley became widely known for the quality of his wines and the attractions of his can-can girls. His advertisement in the *Nashville Times* announced:

> FRENCH CHARLEY
> To the diggers of Nashville
> and the One Mile
> C. BOUEL, T. FAWCETT & CO.
> Novelties Every Night
> at the
> THEATRE ROYAL, ONE MILE
> They will always do their best to please the
> public in general
> THE BAR
> As all the boys know, is always supplied with
> the best of
> Liquors and Drinks
> Knock-me-downs, Pick-me-ups, Smashes,
> Cocktails, Flashes of Lightning, Volcanoes, etc,
> Theatre opens at 7 sharp; ring up in half an
> hour after.

The fashionable theatre of the town was the Varieties, where things were done in the London music hall style, with refreshments brought in. Comic opera, dramatic and vaudeville companies visited.

Shakespeare came to the goldfield by bullock dray. Tommy Hudson was managing a Shakespearian theatrical company at Rockhampton when news of the Nashville rush arrived, and he immediately took ship for Maryborough. There he packed his ladies and their luggage in a bullock dray and, with his future Falstaffs and Hamlets plodding alongside, took the muddy trail to Gympie Creek where Mr and Mrs J. L. Byers, stars of his company, were soon playing Iago and Desdemona in *Othello* to appreciative audiences and to a few astonished Aborigines who watched through a gap in the canvas.

But in spite of such cultural opportunities, the basic fact remained that digging for gold was dry work, and the *Maryborough Chronicle* correspondent on 14 September 1868 was able to illustrate the town's prosperity by listing its hotels.

> Crossing Nash's bridge into the main street, the first hotel which attracts attention is Breddall's, with the sign of "The Maryborough". By hotel, I mean a public house which has accommodation for lodgers. There are large numbers of licensed victuallers' buildings, but few have more than a bar and two or three sitting rooms.
>
> The British ensign floating on a lofty flag staff displays the legend "Golden Age" on a red ground. The quaint, Old English hanging sign swings beneath, emblazoned with the name of the hotel in letters of gold on a blue field. Immediately opposite is that excellent hostelry, the Royal Hotel. Next comes the Prince of Wales, with a theatre attached. Further on, and on the opposite side is Fulton's Melbourne Hotel, and nearly facing it the Newmarket, kept by host Burton, in the rear of whose premises is a livery stable. Close to it is Sinclair's Horse and Jockey.
>
> We now mount the hill patronized by the civil authorities. Here, facing one another, are the two latest improvements—Farley's Mining Exchange, and Foos' Mill. Both the proprietors have recently come from Victoria.

The Carnival Year | 101

Farley's bids to become an immense building. The front bar is one of the largest I have seen in the Colony and adjoining, in a spacious apartment, are two first class billiard tables with every necessary adjunct, not excepting a separate bar. Foos' Mill is an elegant structure and stands on one of the finest business sites in the town.

Thrower's Freemason's Hotel is a little beyond Farley's, and in a separate building is a spacious billiard room and accessories. On the side of the range behind Foos' is Croaker's Northumberland Hotel—a large edifice standing at present in a rather isolated position but, being the booking office of Cobb and Co., is well supported.

Numerous grog shanties were still doing well in spite of all efforts by hotel keepers to have them closed.

October 17: A number of persons were brought before the Police Magistrate at Gympie for sly grog selling, but the informers' testimony proved to be unreliable and the men were discharged. We know how difficult, if not impossible, it is to stop sly grog selling, but it does not follow that shanties should be openly in the main street with the bar and shelves full of liquor, and be seen by the police selling it and not stopped. Shanty-keeping has now assumed such a position that those who have been fined think that they have been unjustly dealt with.

The majority of the diggers did not altogether support the hotel-keepers' stand, and as soon as the court had adjourned the informers were run out of town, lucky to escape with reasonably whole skins.

The goldfields' first race meeting was held on Tuesday and Wednesday, 29 and 30 December 1868. Quite a fair course, considering how hilly the country was, had been cleared near the mouth of Gympie Creek, just up-stream from Widgee Crossing, and first thing on the Tuesday morning it looked as if the whole populaton of the diggings was streaming out along the track leading to it. Along the way were liquor booths already crowded with patrons.

The course looked like a gigantic fair ground set down in the middle of the scrub. Every publican and shanty-keeper in town was there, most of them in sapling stalls thatched with leaves and staffed with attractive young barmaids in carnival costume. A stream of drays—normally used to carry wash dirt to water—ran a shuttle service to keep supplies replenished. As well-primed revellers rolled off to explore the other attractions of the course, nuggets of gold changed hands for a pitch at an Aunt Sally or a toss at the hoop-la. Rouge-et-noir tables were scattered with gold, and even the thimble-and-pea, and three-card tricks found takers.

By the time the racing was due to start, two boys who were trying out the track had collided and been thrown. Both escaped serious injury but one of the horses was killed. Though pronounced "cleared" for the occasion, the circuit included rocky slopes and several patches of timber. The straight was scattered with tree stumps and logs. As many of the riders had been drinking since daylight, riding was reckless and collisions and falls were frequent. Several arms and legs were broken and a number of horses killed.

The first night of the meeting was one huge, roistering barbecue. The second day's racing was a shambles. Spectators crowded onto the track and scattered amid cheers of encouragement as the horses came thundering down on them. Judges' decisions were disputed, argued over and fought over. New Chums who had never ridden a horse in their lives found mounts and before the meeting was over it was estimated that riders had totalled about five hundred. Nobody was killed, but next day Gympie was much quieter than usual.

After that first meeting, the annual race meeting became the goldfield's main social event. It was also recognized as the occasion for the settling of all outstanding differences between Caledonians and Hibernians, and some notable battles were fought.

The year 1868 also saw the building of the Bethel Church at the back of the Northumberland Hotel on Commissioner's Hill. It was constructed of palings, bark, and other building leftovers, and was used by all denominations except the

Catholics who began with a borrowed tent and later had a bark church of their own.

By the time Father Matthew Horan arrived in March, religious feeling was still lukewarm among many members of the community, and the Reverend Father later recalled that as he got off the coach from Maryborough a digger eyed his cassock with some hostility and exclaimed to his mates, "Strike me flamin' pink, here's the bloody parson arrived; it's time for us blokes to get out when the parson comes."

Soon after this a German New Chum named Muller, on the banks of the Mary River, picked up a handful of sand full of mica flakes which glisten in the sun just like gold. Muller excitedly reported that he had found gold in the Mary River, and a man from the *Nashville Times* wrote a humorous story to the effect that gold was so easily got in Nashville that a digger had dived into the Mary River and come up with a handful of it. Most appreciated the joke, but there were men in town at the time who had worked hard and never got a grain of gold, and their sense of humour was wearing thin.

After a session at French Charley's a mob of them, under the leadership of a big digger named MacNamara, decided they ought to pull the newspaper office down. Manager Frank Kidner tried to reason with them without success and managed to escape to seek the aid of Father Horan.

By the time they arrived back at the *Times* office a large crowd had gathered in anticipation of the fun. A long rope had been tied round the top of one of the corner posts, and MacNamara was clearing a track through the press of people so everybody could get a good grip on the rope.

Father Horan addressed the mob, but they took no notice of him. He then spoke to MacNamara, appealing to him as an Irishman and as a Catholic. But by this time MacNamara was feeling like a fight, and the whole thing had become a matter of principle.

He had his men lined up on the rope and was just about to give the order when the priest saw a tomahawk lying handy. With everybody concentrating on the job in hand, no one noticed as he picked it up and took his stand beside the post to which the rope was tied. At last the diggers were ready.

"Now, me boys, a big pull and a strong pull, and pull all together; one, two, three, heave!"

As the strain came on, Father Horan lifted the tomahawk and brought it down across the rope where it went round the post. The rope fell free, the men manning it collapsed in a heap, and some of them were sat on heavily. Fights broke out, and by the time they were settled and reconciliations effected at French Charley's, the newspaper office had received its reprieve.

Chapter Thirteen: Chinaman's Flat

As on almost every goldfield, it was the story of the alluvial gold that was soonest over at Gympie, and among those who pulled out before the end of 1868 was James Nash himself. The *Maryborough Chronicle* correspondent at Gympie reported on 7 September:

> Nash, the original prospector of this goldfield, has sold his claim after working it well for nearly twelve months. It was put up for auction today and knocked down by Mr Booth to a man working for Nash for £100. The claim to any reef struck on it was reserved. The whole of Nash's Gully has been thoroughly riddled, in like manner the flat opposite the site where the first stores went up.

The last of the gold had been taken out of the flat by Chinese diggers who, following their usual custom, had drifted onto the field in small parties behind the first wave of the alluvial rush and, moving onto ground abandoned by less patient European diggers, picked the last of the gold out of it literally a grain at a time.

As often happened, there had been a distraction early in the year which enabled the Chinese to get a foothold.

By May 1868 much of the old alluvial ground was showing signs of being worked out, and reports of gold finds at Kilkivan, Imbil, Yabba, Jimna and other scattered localities within a radius of about fifty miles to the south and west led to several rushes out of Gympie by alluvial miners.

106 | Gympie Gold

On 9 May it was reported from Kilkivan: "There are about 5,000 people on the ground, and a hundred drays loaded with provisions have gone there from Gympie and Maryborough." Chinese promptly moved onto the abandoned ground.

> GYMPIE, May 16: The Mongolians have lately been arriving in gangs. Their favourite place seems to be the flat at the end of Nash's Gully. They are orderly and harmless and we presume there can be no objection to them.

Before long there were about six hundred Chinese on the field. With their usual dogged patience, often working up to their waists in water, they went over every inch of the ground around Nash's Gully and gradually spread out right to the river-bank on one side, and to the alignment of Mary Street on the other. They would have mined the street itself if permitted. Some time earlier, white diggers had pegged out all the lower part of the street, but Commissioner King had warned them off.

Soon the whole area—Chinaman's Flat, it was now called—was honeycombed with their diggings, and from daylight to dark the flat echoed with the click-clack of the cradles. By July the *Nashville Times*—which eventually became the *Gympie Times*—was reporting their activities in a less complacent strain:

> July 4: There are still a few Europeans fossicking in Walker's Gully and parties of miners are at work in Nash's and Sailor's gullies, but most of the claims in the gullies are now abandoned to the Chinese, whose numbers are rapidly increasing. The claims in the bed and on the banks of Deep Creek are still under water.

> July 11: The Mongolians are rapidly usurping sites formerly occupied by Europeans and we understand that one celestial capitalist intends to erect a building on the One Mile for the accommodation of nearly 1,000 of his countrymen who are reported to be on their way to the field. Already they have taken up large claims on worked-out ground and they are making streets on the flat at the foot of Nash's Gully—similar to those formed

by them on the old diggings—on which they can establish their own stores etcetera.

Within a matter of months the Chinese took out of the flat thousands of pounds worth of gold that impatient white diggers had overlooked, and when the original owners came drifting back from the unsuccessful rushes to Kilkivan and other fields, only rubble remained.

"John Chinaman" was always fair game for a "bit of fun" on a goldfield at the best of times, but when things went wrong, as they began to do for some of the alluvial men now, he always became the general scapegoat. There was a "roll-up" and the Chinese were chased off the field. E. B. Kennedy, who was there at the time, described it.

> "Roll up, roll up", we heard roared all through the camp, and at once Celestials were flying helter-skelter, taking flying leaps over claims, sometimes into them, when they would be dragged out by their pigtails and cuffed on again. At first they started laden with buckets, pots, bedding and other gear; gradually this was cast aside as they whirled away with an incessant jabbering which was only equalled by the oaths and shouts of the pursuing party. Those who had coiled up their pigtails got off the easiest, but when that appendage was flying behind, the owner sometimes came to grief, as the waggling tail was too tempting. The Chinese mob eventually outdistanced their pursuers.

The riot, however, had little permanent result. The year was a wet one, and European diggers who forcibly repossessed old claims found working in the water little to their liking and returns insufficient for a European to live on. By September the Chinese were more numerous than ever, attracting the grudging admiration of passing white diggers by their perseverance. A correspondent wrote:

> In passing by today I noticed one fellow getting about half a pennyweight to a dish of the stuff he had just washed, others only a speck. Their camp is very neat, most of their humpies having singular hieroglyphic notices posted outside them. There is a Chinese camp hotel, another inn, a doctor's establishment for the

> stock of Chinese drugs, a butcher's shop, grocers, fruiterers and confectioners, and several storekeepers, together with opium smoking and gambling huts, from which droll music is occasionally to be heard.
>
> Notwithstanding the strong determination at the outset to drive away all Mongolians, there is no question that those who have been permitted to establish themselves have not given cause for reproof, being extremely well conducted, unobtrusive and industrious. Without the Chinese our tables would not be plentifully supplied as they are, with every description of garden produce of the best quality, in profusion and at moderate prices.

The alluvial holes dug by the Chinese were not rectangular like those of the Europeans, but circular. The holes generally went down about fifteen or twenty feet and when abandoned were left open. Unlike the rectangular holes, they rarely caved in, but the top edges crumbled, grass grew thick around them, and they became traps of mud and stinking slime lying in wait for the unwary amid a maze of overgrown tracks and mullock heaps. Horses, dogs and cattle fell into them and were never found. One white digger spent a night in one before managing to scramble out by footholds clawed into the walls with his bare fingers.

By the end of September 1868 most of the alluvial was gone, but there were still occasional rich strikes to keep the hopes of the remaining alluvial men alive. On 23 October the *Nashville Times* reported: "one of the richest finds ever in Gympie, rivalling some of the far famed 'jeweller's shops' of Ballarat".

The find was made on partially worked ground at Walker's Gully. Three months previously, a party of diggers had given up within a few feet of it. The abandoned claim was taken up by an old digger who pitched his tent on it and spent his days scratching about in no great hurry to do much digging, but fond of telling all who came by that his fortune lay under the spot where he normally boiled his billy.

He named his claim the Golden Bar in confident expectation of its richness and continued to take things quietly, doing no more digging than he felt inclined to do until one day, down in the old shaft under the spot where he had been boiling his billy, he put his pick into a nest of golden nuggets.

From all around diggers came running as he brought the first big lumps of gold to the surface. First reports claimed he had at least two hundredweight of gold; later on the figure rose to a ton. At the end of the day there was one large nugget weighing about 700 ounces and a heap of smaller ones ranging from about 10 ounces to 150 ounces. Dirt washed that day had yielded from 70 ounces to 100 ounces to the dish. Within four days the yield totalled more than 4,000 ounces.

There was a rush to peg all the surrounding ground, but little gold was found in it.

A few more big nuggets and rich patches of alluvial came to light in succeeding years, but they were isolated finds. Gympie's alluvial boom was over within a year of Nash's first find. From then on the field belonged to the reefers.

Gympie's total alluvial gold production has never been calculated. Of that found in the last months of 1867 when the richest of the shallow ground was being worked, there is no record at all. Men kept quiet about what they got and generally took it away with as little fuss as possible.

From January 1868 there are gold escort returns showing that a total of 84,792 ounces were taken from the field by escort in 1868, but much more was taken away by miners who either objected to paying heavy escort charges or felt they could guard it better themselves. Early escort figures for gold taken out in 1868 were: January 3,526 ounces, February 4,968 ounces, March 4,409 ounces, April 8,454 ounces, May 10,918 ounces. As the first crushing battery came into operation in April, a large part of the escort gold from then on would have come from the reefs.

One estimate of the total amount of alluvial gold that came from the creeks and gullies of Gympie during the twelve-month boom period puts the figure at 150,000 ounces. Whatever the correct figure may have been, the greater part of it was probably taken during the first few months when the shallow ground was being worked.

After Nash sold his original claim in September it was widely believed that he, his brother John, Billy Malcolm and

Leishman between them had taken gold worth about £30,000 out of their ground during the twelve months they had worked it. Nash admitted to having cleared £7,000 after paying all expenses.

When one takes into account the fact that £7,000 would have had a purchasing power—on today's standards—of something like $100,000, it becomes a mystery what Nash did with the money. In one way or another he apparently lost it all. He was married in Maryborough on 6 July 1868 to Catherine Murphy. He made a number of investments in businesses ranging from gold mines to a drapery store, and apparently lost money on all of them. He was no doubt an easy mark for diggers down on their luck.

He claimed the Government's £3,000 reward for his discovery of the goldfield, but though the Gympie gold had probably saved the colony from bankruptcy, the Government was reluctant to pay the reward, claiming Nash was not entitled to it because the field was not within the specified ninety miles of Brisbane. After considerable argument £1,000 was paid and Nash and his wife left on a trip to England.

They returned later to the Gympie district and raised a family of two sons and a daughter. But Nash had no part in the reefing boom his goldfield was by then experiencing. Eventually, in 1885, the Government appointed him keeper of the powder magazine at Gympie on a salary of £100 a year. He became a local identity, fond of talking about the good old days and tracing the outlines of the old alluvial gullies fast disappearing under the paved streets and substantial buildings of the growing town. He died on 5 October 1913, aged 79, and was buried in the Gympie cemetery.

Chapter Fourteen:
Breaking out the Quartz

The reefs of Gympie were discovered so rapidly after the opening of the Lady Mary at the beginning of November 1867 that by the end of February 1868 the bank had all the rich specimens it could store, boxes and bags were full of them, bunks covered hoards of them, and the local correspondent of the *Queenslander* was writing: "In the name of common sense, why are not crushing machines sent up; I know of reefs rarely exceeded in richness."

Many of the reef claims were so rich, in fact, that their owners were crushing the quartz by hand in iron dollies and getting returns as good as those of the men on the best alluvial ground.

The Pollocks and Lawrence, feeling they had more on their hands than they could manage alone, attempted to float the Lady Mary P.C. (Prospector's Claim) as a company in Brisbane with a capital of £12,000 in 4,000 shares of £3 each, five-twelfths of which, fully paid up, were to go to the owners. They found men who had never seen the field still wary of sinking capital in it. There was a good deal of dickering, and when the owners refused to give an undertaking that they would not sell their shares until after the first crushing, the capitalists took fright and the proposal fell through.

The Pollocks and Lawrence slogged on alone and the phenomenally rich ore of the Lady Mary P.C. yielded them a fortune.

The reefs generally ran at an incline and often had branches, or leads, running out from them. If one followed a lead it might show the way to the main reef. On the other hand, earth movements in the past might have broken it so that a miner following it down would suddenly come to an end, with only barren rock beyond. Some leads were very rich.

The story was told of a digger's wife who was bringing dinner to her husband and his mates at their claim when she nearly tripped over a piece of quartz and, on picking it up, found the under side of it was almost pure gold. She told the men who, forgetting all about their dinner, came trooping back to the spot with their picks and straight away started digging. They soon found the lead the specimen had come from, and every inch of the quartz was laced with gold.

All the rest of that day they dug on, and after it was dark the digger's wife lit a fire and cooked them a meal. Even then the digging did not stop; one man ate while the others continued to dig. By daylight they were all rich.

From all over the field men flocked to peg claims around them. None found gold. Then the lead cut out and not a trace of quartz remained. But one night's work had brought four men fortunes.

One of the problems reefers faced was that the line of a reef was not always predictable. A fault—or break in the level of the rocks—could chop it off short and take the remaining part of it down to a different level altogether. Reefs could change direction, divide into two, or link up with others in an elaborate network of underground rock that was impossible to follow from the surface.

When a reef was discovered the Gold Commissioner plotted the line of it, or laid it down. Claims were granted on either side of the prospector's claim along the line of the reef and numbered to distinguish them. But if a reef struck a fault, or swung off its expected direction, a man who pegged ground where the reef should have been could find himself digging in barren rock, while another might find himself with two reefs on his claim.

This happened with the Lady Mary and the Caledonian reefs, which were originally discovered running roughly parallel to each other.

Acting Gold Commissioner Davidson had laid down the line of both the Lady Mary and the Caledonian reefs, but as prospecting along them continued it was found that the Lady Mary did not follow the line, and the reef was lost for several months. Then it was discovered that the Lady Mary and Caledonian reefs, after running side by side for some distance, joined up to become one. Prospectors who had been granted claims along the supposed lines of two separate reefs suddenly found themselves with only one reef between them, and came storming to Gold Commissioner King, who had by then taken over, to seek a solution of the tangle.

King re-plotted both reefs, laid down the combined reef from the point where they joined, and began hearing the claims of dozens of furious miners who had been digging in the wrong place. Some had abandoned their original claims and taken up new ground, others met in head-on confrontation, Lady Mary and Caledonian claim-holders each asserting their right to the same section of the combined reef. King ruled that as the Lady Mary claims were the oldest, holders of such claims should take the disputed parts of the combined reef. Many gained or lost fortunes on the decision.

A few lucky ones found that their claims included both the Lady Mary and the Caledonian reefs at a point before the two united. Among them were the holders of the Caledonian P.C., Messrs Goodchap, Kift and Morgan, and the holders of the Lady Mary P.C., Messrs Robert and Alexander Pollock and Lawrence. The latter party, after opening up the Lady Mary claim, went on to prospect their section of the Caledonian reef and, as luck would have it, put down their shaft right over the top of a small "golden patch". Fellow reefer Nugent Wade-Brown, who was taken down to have a look at it, wrote:

> On the north and south end of the shaft the quartz bristled with gold. They went down about thirty-five feet and there the gold ceased. They then started to break down on the north and south ends of the shaft and discovered that the gold did not continue beyond about two feet each way. Down to the thirty-five foot level they had won about 4,000 ounces of gold.

It was a time when anything could happen. Digger William Clarke later recalled:

The red flags seemed to be flying everywhere in those days. Old White, from Maryborough, had just got from one hole in White's Gully as much gold as he and two mates could comfortably swag up to the bank.

A party of four just in from New Zealand [Gilbert Muir, Adam Black, Robert Black and Robert Drew] pitched their camp on the edge of a dense scrub, just off the mailman's track. Surface quartz was indicative of a reef just below. The four Maorilanders bullocked into the quartz in orthodox fashion and sank fifteen feet on the quartz quarry without a halt. Just below that depth a complete transformation of appearance set in, with buckets of lumps of solid gold set in the quartz. From this claim the four of them lifted fortunes of many thousands of pounds before selling out.

Three other New Zealanders, Harry Jackson, Bill Freeman and another, brought in 500 ounces of gold picked out at a depth of fifteen feet from a reef at the Two Mile. A week after that Bill Harvey excavated from the London six buckets from one breaking down that were heavier with gold than with stone.

The goldfield's first quartz-crushing machine, the Pioneer, arrived on Sunday, 8 March 1868, and work was begun erecting it up-stream from Nash's Gully at what was called the First Pocket. It was small and primitive, consisting of two batteries of five stamping heads each, and driven by a small, portable, twelve-horsepower steam engine. But to the impatient diggers who gathered daily to see how work progressed, it was the most important thing on the field, and its completion the most urgent.

On 29 April, before the job was finished, they induced the owners, Pye, Bachelor and Co., to give one battery a trial run with three hundredweight of ore each from No. 1 North Caledonian, and Caledonian P.C. It was picked quartz which had been a constant worry to its owners because it was too bulky to lodge in the bank, but so rich that it was always in danger of being stolen. The three hundredweight from No. 1 North Caledonian yielded 450 ounces of gold, that from the prospector's claim, 367 ounces—rich enough to send every reefer on the field digging with renewed enthusiasm.

The method of working the stampers was to feed the ore into a sort of iron box in which the five stampers operated, and wash the resulting powder through with a stream of water. The mixture flowed out over copper plates covered with mercury—or quicksilver—which formed an amalgam with the gold and held it on the plates. The amalgam was then scraped off, heated in a retort to drive off the mercury, and the gold remained. The mercury was condensed and collected for re-use.

An early problem with the Pioneer mill was the lack of adequate water pumping gear, so that until July the machine operated on only one battery of stampers. With the mill putting through no more than about thirty tons a week, and rich stone being brought to the surface every day, huge stockpiles of ore built up and the waiting list lengthened daily.

Much of the ore that went through in those early crushings was specially selected stone which had been lodged in the bank for safekeeping. Every lump of it was fed individually into the stampers and washed through over copper plates constantly replenished with quicksilver to make sure none of the gold was lost. But even among larger lots, some of the yields were sensational. Among them were: Caledonian P.C., 708 ounces of gold from thirty-nine tons of ore; 1,188 ounces from thirty-eight tons; 1,278 ounces from forty-two tons; Lady Mary P.C., 1,358 ounces from seven hundredweight; South Lady Mary, 1,167 ounces from twenty tons.

On 2 July 1868 the Queensland Government Geologist for South Queensland, C. D'Oyly H. Aplin, reported:

> The gold, though generally under standard fineness in quality, occurs in unexceptionally heavy masses in the quartz veins. In some instances, as in Dodd's Reef especially, it has been obtained in broad plates an eighth of an inch or more in thickness. In the Caledonian Reef (Lawrence and Pollock's claim) it has been found in bundles of filamentous threads, united together with ribbon-like lengths and bent into not ungraceful loops and curves. In other instances it is in amorphous, ragged masses of great size and weight, but I have not yet met with or heard of it in a crystalline form.

> For exceeding richness in patches of limited extent, the quartz reefs of Gympie have, perhaps, never been surpassed, if equalled, by those of any other goldfield, parcels of from six to eight hundredweight having yielded from one-tenth to one-sixth of their weight as gold. Large quantities of gold have been obtained by hand crushing in a mortar.

With reports of the rich crushings reaching Sydney and Melbourne, the Canoona bogy was finally dispelled and experienced reef miners from the southern goldfields began heading north to form a third wave of the Gympie rush. On 16 July 1868 the *Maryborough Chronicle* reported:

> The screw steamer *Hero*, the first vessel direct from Melbourne, reached Hervey Bay on Tuesday (14 July) with 320 passengers, all bound for the Mary River goldfield. Owing to the deep draught of the *Hero*—sixteen feet—it was thought advisable not to attempt to bring her up the river with her passengers. So the steam tender *Hawk* was engaged to bring them up, and six of the *Hero's* boats were filled and towed by her. Passing the wharf the passengers in the steamer and in the boats set up a vigorous cheer, and then the boats threw off and diggers in them lost no time getting ashore.

Though Maryborough's wharves were lined with ships and a second quartz-crushing machine lay waiting to be taken to the field, most of the town's houses were empty, all the inhabitants having gone to the diggings. Captains of steamers, knowing there would be no local labour to unload their cargoes at Maryborough, picked up Aborigines from Fraser Island, opposite the mouth of the river, brought them up to the town to do the job, and dropped them off at the island again on the return voyage.

On 18 July the *Chronicle* reported:

> From Gympie we hear that new reefs are almost daily being discovered, whilst the old ones are increasing in value.
> The *James Paterson* arrived with 110 passengers yesterday. The *Florence Irving* left Sydney on Wednesday (15 July) and the *Black Swan* on Thursday, both full. The *Saxonia* was expected to leave yesterday (17 July).

One-time convict James Davis, known to the Gympie Aborigines, among whom he lived for many years, as Duramboi. (From Tom Petrie's *Reminiscences of Early Queensland*.)

James Nash, English migrant prospector, who discovered the Gympie goldfield. (From *Queensland 1900*.)

ABOVE: Much of Gympie's alluvial gold came from shallow gravels along the gullies. (Richard Daintree photograph.)

BELOW: Cradling for alluvial in a scrub-surrounded gully. (Richard Daintree photograph.)

ABOVE: Many miners covered their shafts with rough bough shelters to keep the sun off the man on the windlass. (Richard Daintree photograph.)

BELOW: Beside every gully where gold was found, clusters of tents and rough bark shanties soon sprang up. (Richard Daintree photograph.)

ABOVE: Mullock heaps soon rose among the tall timbers as shafts got deeper. (Richard Daintree photograph.)

BELOW: Gympie at the height of the rush during the 1870s. (Richard Daintree photograph.)

ABOVE: By the end of 1868 Gympie consisted mainly of two rows of bark huts lining Mary Street. (From the *Australasian Sketcher*.)

BELOW: The start of the 1870s saw Mary Street, still an unpaved, winding track, flanked by shingle-rooved stores backed by the bush. (From the Gympie Centenary Publication, 1967.)

Diggers struggling to keep going on mines that did not pay often displayed a dry humour that caught the eye of the cartoonist. (From *Travel and Adventures in Northern Queensland*.)

As the mines went deeper the man on the windlass was replaced by a horse-powered whip to pull up the buckets of ore. (From *The Never Never Land*.)

ABOVE: Before the introduction of steam power the most widely used method of raising ore from the mines was by means of the whim, consisting of a large drum round which the rope was wound by a horse walking in a circle. (Richard Daintree photograph.)

BELOW: Gympie under water during the great flood of 1893. In the foreground is the pit-head and whim of the No. 2 and 3 South New Zealand mine. (From *The Queenslander.*)

Gold-miner William Couldrey, who made and lost several fortunes on the Gympie goldfield and retired to Sydney a wealthy man. (From *Queensland 1900*.)

Scottish migrant Matthew Laird, who developed a mine nobody wanted into Gympie's top gold producer. (From *Queensland 1900*.)

ABOVE: All that remains today of the once famous Scottish Gympie mine is the old retort house where many of the great golden cakes of the boom days were produced.

BELOW: Victoria House, on Redcliffe Peninsula overlooking Moreton Bay, was built by retired gold-miner Jacob Pearen, who liked to watch the ships passing.

ABOVE: Gympie's last big mullock heap, in the once famous One Mile area, is rapidly disappearing as stone is taken to level a playing field.

BELOW: Stampers, pit cages and an old boiler from the No. 2 South Great Eastern mine now stand in the grounds of the Gympie Museum.

ABOVE: From the balcony of the old Freemason's Hotel today one looks down Mary Street to the spot where the road bends where James Nash had his prospecting claim.

BELOW: Gympie's Town Hall, just below the ridge where the famous "Mother of Gold" was found, is flanked by a monument to James Nash, discoverer of the goldfield.

ABOVE: Inconspicuous among more elaborate monuments, James Nash's grave at Gympie Cemetery is marked by a bronze plaque stating simply: "In memory of James Nash who discovered the Gympie Goldfield 16th October 1867. Born at Beanacre, Wiltshire, England, 5th September, 1834. Died at Gympie 5th October 1913."

BELOW: Gympie's Memorial Park, with a monument to James Nash, now covers the gully where he found the first gold.

RIGHT: Owners of Gympie's new El Dorado mine, whose shaft goes down beside the Bruce Highway, main route to North Queensland, seek a new reef below Albert Park, the town's main playing area.

BELOW: Phoenix Reborn mine, taking its name from one of the famous mines of Gympie's gold boom days, goes down today on a vacant allotment in Ray Street.

> The *Blackbird* left Melbourne last Saturday (11 July) and the *Omeo* was to follow. The A.S.N. steamers to Brisbane are also well filled with miners making their way to this district.
>
> Over 1,000 persons have landed in Maryborough during the last fortnight bound for the diggings.
>
> Some little difficulty and delay has been experienced in getting the huge boilers for the Enterprise Quartz Crushing Company to Gympie on account of their great weight. A carrier named Hendry, who has had considerable experience in conveying quartz machinery in Victoria, has undertaken the job. Yesterday one of the boilers was hoisted safely on the top of a four-wheel timber carriage and started on its journey. If the weather keeps fine and the roads firm, the boiler will be on the diggings in a few days.

The boiler arrived safely, and before long the field's second crushing machine, the Enterprise, was operating at the Second Pocket, half a mile further up-river from the Pioneer.

Teamsters' freights from Maryborough to Gympie were anything from five to ten pounds a ton, depending on the weather. When it was wet there were stretches of track where the mud was feet deep, and what was normally a two-day trip sometimes stretched out into several weeks.

It was about this time that a bullock teamster, loaded with liquor from Maryborough to Gympie, had one of those trips when everything goes wrong. The grass was poor and so were his bullocks. A couple of them died of pleuro and several others looked like following suit. His harness kept breaking and his load kept slipping.

At the edge of Black Swamp, on the Gympie side of Tiaro, he decided he had had enough. He pulled the wagon into the shade, unyoked his bullocks, threw down his hat and cracked a bottle of Scotch. The first man past was invited to join him in a drink, did so, and at last moved on. Others followed, and while the Gympie publican sweated and fumed for his consignment, the bullocky kept open house by the roadside, allowing no man past without a drink. Fred Eaton, who was carrying between Gympie and Maryborough at the time, passed him about a dozen times on his trips to and fro.

Word of what had happened soon got through to the publican, but it was more than a fortnight before he could hire another teamster to go out and bring in what was left of his consignment. By the time the two carriers had had a few drinks together under the trees there was not enough liquor left to be worth taking to Gympie, so both of them headed back to Maryborough.

Apart from the crushers, there was still no heavy machinery on the field. Each mine shaft consisted of little more than a staging of logs built up about ten or twelve feet above ground level to give room for the excavated stone to be tipped out, and something to raise the buckets—either a hand-operated windlass or a horse-driven whip or whim.

The whip was a sort of derrick at the top of which was a pulley over which the rope was passed. The horse was harnessed to the rope and pulled up the wooden or greenhide buckets by being walked away from the shaft. In the whim the rope was wound on a drum which was rotated by a horse harnessed to a pole and walking in a circle.

The old whim horses were trained to obey instantly, and they got so used to their work that when taking the buckets away from the bottom of the shaft, they would move off very gently until the bucket was clear of obstacles, and then make the pace until it was a foot or two above the floor of the landing. Whip horses soon wore a track in length equal to the depth of the shaft and learnt to stop at the end of it. There they would wait until the bucket was emptied and the man at the pit head called the order, "Turn", at which they turned and walked back to the shaft, thus lowering the bucket. If a whip or whim horse did not stop in time the bucket of ore would hit the pulley, and generally break the rope and go crashing back down the shaft to the considerable danger of anybody below. A horse was no good unless he obeyed on the instant.

One of the miners used his whip horse on Saturdays as a saddle horse to ride in to the bank to collect the men's pay. Small boys were in the habit of hiding behind a tree and, as the horse was passing, calling out "Turn", roaring with laughter as the horse obeyed with such alacrity as almost to unseat the rider.

The sudden expansion of reefing activity on the field meant a run on explosives for blasting. One storekeeper sold his whole stock of eight hundred pounds of gunpowder in three days, and by the beginning of July there was not a pound of powder or a length of fuse to be had at any price. One of the reefers wrote: "It was like work down in those straight shafts; no powder or fuse on the field. Stripped to the waist, we put one foot in the noose of a rope, took a double handed hold of the lowering rope overhead, and were lowered down. We worked with hammer and gadge at the cleavage in the diorite, the perspiration running down our naked bodies like water."

There were, however, plenty on the field to envy them. "Old Bendigo men pasing would look down the shaft and call out, 'You stick to that mate; better than all the 'luvial'."

The long-awaited gunpowder arrived early in September, without incident, but causing considerable alarm along the track. A dray loaded with about two tons of it in uncovered barrels stood outside the *Maryborough Chronicle* office for half an hour or so while the driver and bystanders yarned, lighted their pipes and tossed the matches beneath it. The powder also survived the two-day trip to the diggings, still uncovered, was pounced on by waiting miners and soon "reports of the blasting sounded like the cannonading of bombardment".

A miner was killed and his mate badly injured when tamping a charge of powder into the drill hole with a rod of iron instead of wood or copper. A spark from the iron on stone ignited the powder and caused an explosion that sent a shower of stones cascading out of the shaft like a volcano. Both men, though terribly injured, grabbed the rope and were hauled to the surface, where the one died of his injuries.

Another miner went down his shaft after a quarter of an hour to see why a blast had not gone off. He called to his mate to send down a candle and soon after there was an explosion. The man was brought up badly injured. He did not know what happened. The candle may have ignited loose gunpowder which set off the charge, or the original fuse may have been smouldering all the time.

Two men attempted to fire a shot that missed twice. They then began drilling out the charge, which exploded. Both were badly injured.

A new doctor arrived in town with an apprentice, had himself a comfortable slab house built, and put up on the front wall beside the door a brass plate setting out his qualifications. Many of the diggers had never seen such a thing before, but they were prepared to give the new man the benefit of the doubt until one day a digger had an accident down his shaft and was hauled to the surface with his leg broken in two places. While his mates put up a shelter of branches to keep the sun off and made him as comfortable as possible, a lad went running for the new doctor who happened to be nearest. The boy soon came panting back to announce:

"The doctor won't come unless he's paid first."

For a moment the diggers stared in disbelief, then an angry growl went up.

"He won't, won't he? Roll up boys!"

"Roll up, roll up," went the call, and every digger who heard it dropped his tools, passed on the word, and came running. A few waited with the injured man; the rest of them and their dogs went roaring up the hill to the brass-plated house.

The doctor saw them coming, panicked, snatched up his loaded shot-gun and fired. Luckily for him the range was too great for the shot to do more than sting a few of the leaders, but that was enough. Revolvers were drawn and shots flew wildly as the diggers charged—in far more danger of wounding each other than their quarry.

They smashed through the door like an avalanche, only to find the doctor had bolted out the back. The building's only occupant was the apprentice, considerably alarmed, but determined, apparently, to hold his ground until he found out what it was all about. With a hundred or more men pushing round to explain, the apprentice made no talk about fees, but seized his absent master's instruments, and accompanied the miners back to the injured man. He did such a good job on the broken leg that he was well paid and became "Doc" from then on.

Breaking out the Quartz | 121

Accidents with gunpowder were common on the field, and many miners lost a limb or were blinded by premature blasts. Jim Hayes, who kept a hotel at the One Mile in later days, had lost both his arms, one leg and an eye in a mine explosion. He made light of his disabilities and could pick up a glass of beer with the stumps of his arms, drain it and put it back on the counter without any difficulty.

In spite of fatalities and injuries, gunpowder continued to be handled very casually on the goldfield—until about four o'clock one morning early in 1877 when the whole town was jolted to wakefulness by a terrific explosion that shook the place as though an earthquake had hit it. Townsmen turned out to find the middle of Mary Street spouting flames which lit the slopes of all the surrounding hills while further explosions sent timber, harness and all kinds of merchandise erupting into the air as though from a volcano.

The fire was centred on the store of Messrs Woodrow and Scott who traded in all kinds of mining goods, including gunpowder. They had their main magazine out of town, but always kept a good many barrels of powder in the store to meet immediate trading requirements. These went up one after the other as the fire reached them.

The only fire-fighting equipment in town was a few square, steel ship's tanks set up at strategic points along Mary Street, kept constantly filled with water, and a row of buckets hung above each. A few grabbed buckets and filled them, but none got within a hundred feet of the fire. One man was hit by a piece of flying timber and killed.

The fire came at the time of the "Russian scare" when it was widely believed that a Russian invasion of Australia was imminent. As the second barrel of powder went up, one citizen came tearing out of his shanty in his nightshirt yelling at the top of his voice, "Run for the mountains; the Roosians are coming." Behind him stumbled his buxom wife, pleading piteously, "Charlie, me husband, come back; would you run away and leave me to they barbarians?"

A large part of Mary Street was wrecked, but the fire at last stirred the Government to build a brick powder magazine out of the town and introduce strict regulations covering the use of gunpowder.

122 | Gympie Gold

Many of Gympie's best-known mines were opened during the year that followed Nash's original discovery. William Clarke recalled:

> Someone got over the Deep Creek, went down on a reef, and abandoned it, leaving the windlass standing. A new party got onto the hole and got gold at once. The whole line was taken up and named the New Monkland. Five claims were pegged out. Murphy's party discovered unregistered ground between numbers three and four and wedged into it. Results proved it to be the richest claim on the line.
>
> A digger known as Humpy Back Con with five mates had been quietly digging down to the quartz at the One Mile [Afterwards called the Phoenix], and their prospector's claim and the Number One took out some good returns.
>
> On the south side of the river, at Jones Hill, the Otago opened on rich gold. The Macphersons [brothers of the bushranger known as the Wild Scotchman] opened the True Blue.
>
> On the Smithfield line a singular mining freak had occurred. A party of German miners, some of them educated in the Saxon school of mines, opened No. 1 North. The reef took a jump down perpendicularly and was lost for years until the continuation was got accidentally in a reef half a mile away, richer than ever, and the Great Smithfield was started.

There were men in those days who literally threw gold away on their mullock heaps and walked off their claims in disgust, leaving them for others to take over and mine for fortunes.

William Couldrey, Nugent Wade-Brown and Fred and Robert Lord, who were in partnership at the time, were looking over some of the old mullock heaps one morning after rain when one of them was attracted to the top of an old shaft on the Lady Mary reef. There was dull, brown rock all round them except at one spot, apparently where the previous owner had tipped out his last bucket of dirt. At that spot specks of gold could be seen gleaming in the sun.

The partners lost no time in moving in. They rigged up a new windlass, cleared an accumulation of rubbish out of the old shaft and went to work to raise some stone.

Couldrey took the first shift on top, winding up each greenhide bucket of ore and carefully watching the contents of each for any tell-tale gleam of yellow as he spilled it out on the ground. Several buckets came up and were emptied without disclosing any trace of it. Then he emptied another bucket and saw what he was looking for. He crossed to the mouth of the shaft, knelt at the mouth of it, and called down to the others.

"There's gold in the last bucket you sent up."

There was a tense wait then, as the men below did some frantic digging. Then, suddenly, there came a violent shaking on the rope to signal that they wanted to talk to him. Couldrey bent over the shaft.

"What is it?"

"Come down; come down quick; we've struck a bloody jeweller's shop."

Couldrey took hold of the rope and lowered himself down. A jeweller's shop was right. There was more gold than quartz on the floor of the shaft. There was so much gold that the scattered pieces of quartz had to be torn free from the reef with the point of a pick. Said Wade-Brown later:

> We made a strong deal box which was placed in my humpy to keep the richest specimens in. At one time there must have been £3,000 worth of gold in it. Our first crushing was at the Pioneer mill—some three tons of specimens—and the result was phenomenal. A stone we sent to Sydney for exhibition, about nine inches long and six inches deep, yielded, when smelted, 396 ounces of gold.

Couldrey was consistently lucky and was a good businessman as well. He came to Australia at the age of twenty-nine with a lung ailment, but cured it in the course of one long droving trip, and arrived at Gympie in January 1868. He used his early gains shrewdly, led the way in introducing machinery to the field, invested in sound mining ventures and eventually owned shares in almost every notable mine on the field, as well as having many outside interests.

On 28 September 1868 the goldfield's most powerful ore-crushing machine to date, the Victoria mill, began operation on the banks of the Mary River between Walker's Gully and Deep Creek, with a bottle of champagne smashed on its flywheel and a hogshead of beer among all present. The surrounding country was a network of new reefs, and the landscape a mass of stockpiled stone. The correspondent was enthusiastic.

> The engine is of thirty horsepower, driving three batteries of five stampers each, and the weight of each stamper being seven and a half hundredweight. The stampers will be fed by hand, each battery being boxed off so that three different lots of quartz can be crushed at the same time. The gold saving appliances combine the very latest improvement of copper plates and deep ripples on the upper table, and blanketing on the lower table. The blanketing is regularly lifted without loss of time or inconvenience. An amalgamating barrel is also provided, as well as a complete and commodious retorting furnace. No trouble or expense has been spared to make the gold saving appliances perfect in the most trifling detail.

The new machine did much to relieve the pressure of the stockpiles of ore which had been building up steadily throughout the year. But by then reef claims extended more than six miles beyond the first claims that had been opened, and cartage of ore to the crusher by horse dray from the distant claims ran as high as ten shillings a load and more. In addition to that, there was the charge of twelve shillings and sixpence a ton for crushing the ore.

As mines became deeper they needed machinery to work them, and this was beyond the resources of most miners. Those who were doing well began to form companies, not only to work their own claims, but to explore new ground. Many working miners became shareholders in other claims; many others, on good ground but without capital, were forced to sell out and go to work for wages—about two pounds ten shillings a week and no compensation for accidents except what might come from public subscription.

New ground, where the reefs were less accessible, needed capital to develop it. Outside speculators moved in, and the market was flooded with reefing shares in claims that had never been worked.

A correspondent writing from Gympie in August asked the press to warn the public against many of the Gympie mining companies being advertised in Brisbane. He quoted one whose five-shilling paid-up shares were at twenty shillings and another whose shares were also at a premium, in which not a pick had been put in the ground. These, of course, would explode, he said, but by then the originators would have made many hundreds of pounds by selling the shares. The real sufferers would be the genuine concerns on which hundreds had been spent by the original prospectors who were now forced to sell shares because they had no money left.

This was, in fact, what happened. Many investors in mines they had never seen lost their money, and the market for shares dried up—for good mines and duffers alike, and in difficult days to come it was mainly the locally owned companies that kept Gympie going.

Chapter Fifteen: Bail Up!

The rush that followed reports of the Mother of Gold and the Curtis Nugget brought to Gympie not only a flood of experienced miners, but a motley rabble of boom town parasites, thieves and robbers of every sort, and the easy-going times of the opening months gave way to a wave of violence. Early arrivals from southern goldfields had expressed amazement that though Gympie diggers were hoarding gold worth thousands of pounds, few of them carried firearms. They began to need them now.

It had been rumoured on the field that Nash kept a large quantity of gold in a camp oven buried in the dirt floor of his humpy at Nash's Gully. He knew the men working around him and apparently thought it was as safe there as anywhere else. Then, one night as he came back to his camp he was set on by a burly tough who demanded to know where his gold was. Nash insisted it was all in the bank. The other refused to believe him, and Nash was being beaten up badly when an American Negro came charging onto the scene, dragged the attacker off, and hurled him bodily into the bush. The man was never seen around Gympie again, but Nash, like many others, bought himself a revolver.

Fighting in Mary Street increased threefold. In the early days men regarded a fight as the normal way to settle a difference and there were no hard feelings afterwards. An appeal to the law—except in the case of a mining dispute—was regarded as unmanly. Now there was a growing tendency for the old Nashville men to close their ranks against the newcomers—"scrubbers" they were called—and fights became more vicious.

A scrubber would often start a fight to provide a distraction while his mates picked pockets or broke in through the back of flimsily built shops. The risks were considerable. If spotted, those who were grabbed by the police were lucky. Others were bashed insensible by angry miners and dragged into the bush to regain consciousness or not as fate decreed. A man who came round aching and stiff from the bruises of many fists, and maybe with a broken bone or two, had a long, painful walk back to Maryborough ahead of him. To show his face on the field again would be suicide.

The women of Gympie were still a small minority. Brawls over barmaids and dance-hall girls became more frequent, and it was not uncommon for a man who had drunk too much, or got himself knocked out in a fight, to wake up in the morning and find that his wife, or girl, had gone off with somebody else.

A man was found dead in his tent, a bullet-hole in the canvas corresponding with another in his skull showing he had been shot from outside. Suspicion settled on his wife who was notoriously fond of other company, but nothing was ever proved.

Another digger was found in the scrub with a pick driven through his skull, pinning him to the ground. There was gold still in his pockets, so the motive, once again, was probably a quarrel about a woman. The murderer was never discovered.

No man, if he was wise, now ventured out at night without firearms. One protested bitterly in a letter to the *Nashville Times*.

> Surely the authorities must be anxious to encourage crime and outrage, otherwise they would never leave the inhabitants of the One Mile without some protection. They send two police officers to this place on Saturday night, but their arrival is a signal for the assembling of all the rowdy elements in the district, so much so indeed, that a man unarmed would not be safe to pass along the street up to four o'clock in the morning.

Only a few mornings ago, about three or four o'clock, having occasion to go some distance on important business, I was followed by a horde of these midnight ruffians.

I would, I have no doubt, have been severely dealt with were it not for the fact of having in my pocket a six-barrelled speaking trumpet, and presented it at the head of the first vagabond who came, waddy in hand, to knock me down. When I told him that if he advanced a single foot I would blow off the roof of his skull to where he would require something more than a lucifer to find it, the whole crowd turned away, no doubt much dissatisfied at finding one prepared to meet them.

The first gold escort, consisting of a coach and four mounted troopers, had arrived at Maryborough from the diggings on 29 January 1868 with 3,525 ounces of gold. "The reason more gold was not brought down," reported the *Maryborough Chronicle*, "is that the diggers refuse to pay sixpence an ounce escort fees, considering that, for the short distance the gold has to be conveyed and that the Government runs only a common carrier's risk, threepence per ounce should be ample payment."

From then on escorts ran regularly, but diggers maintained their stand and consignments were fairly light. Men with gold preferred to take the risk of bringing it out themselves, travelling alone at night, or by La Barte and Co.'s coach which ran regularly between Gympie and Maryborough three times a week. Many were saying it could be only a matter of time before somebody was robbed.

Predictions were fulfilled about 7 a.m. on 6 April 1868 when the coach, with thirteen passengers aboard, was making slow progress behind straining horses up a rise known as Scrubby Pinch, nearly three miles out of Gympie.

"Right-oh, bail up!"

The coach horses stopped of their own accord and the driver instinctively jammed on the brakes to prevent the coach running back. Three men, masked with handkerchiefs, strode out from the trees. The leader, a well-built man in Californian hat and blue blanket worn poncho fashion, gestured meaningly towards the passengers with a double-barrelled gun he was carrying.

"Get down and shell out quick."

As the passengers piled out, the second man, something of a dandy in Crimean shirt, moleskins and slouch hat, kept the driver covered with a revolver, while the third, smaller than the others, raggedly dressed, and trembling with nervousness, thrust an ancient horse pistol he was carrying into his belt and came forward to search them.

He did not make much of a job of it. He fumbled while trying to pull what was apparently a passenger's handkerchief from his pocket, and then decided to leave it there. The supposed handkerchief was, in fact, a bank bag containing two hundred pounds in notes.

Another passenger surreptitiously dropped his purse, which contained forty pounds, on the ground and put his foot on it. The searcher missed that too.

The clumsy searching took more than twenty minutes and yielded a total haul of only fourteen ounces of gold, thirty gold sovereigns, and £250 in banknotes. Then the bushrangers stood back and kept the coach covered as the passengers climbed aboard and the driver whipped up his horses. By the time the driver was able to report the hold-up to the police in Maryborough it was to be presumed that the robbers were back in Gympie drinking their spoils.

Visiting Gympie at the time of the robbery was the Rockhampton manager of the Bank of New South Wales, R. H. D. White, who had come to establish a branch on the goldfield—to the great satisfaction of the diggers, who were still dissatisfied with the poor price paid for gold by the Commercial Bank and hoped the competition would improve matters.

With the new bank established and Mr C. J. Buckland installed as manager, White took the coach for Maryborough on 19 April. Buckland accompanied him, and it was rumoured that they carried a large quantity of gold which had been hoarded on the field for a long time and sold to White because of the better price he offered.

Roads were heavy from rain, travelling was slow, and the coach pulled up for the night at the Currie Hotel, about twelve miles from the diggings. The *Maryborough Chronicle*'s report of what followed was taken from White's own story.

> Five minutes after their arrival at Mr Booker's Currie Hotel the two bank managers were sitting on a sofa taking a glass of wine together when a scream was heard, and in rushed a man lodging at the house, and after him came a masked man having a large revolver in his hand, who ordered them to bail up.
>
> Mr White had just taken off his coat, but had on a belt in which was a large Tranter revolver and another Smith and Wesson six shooter, copper cartridge revolver, five chambers only of which were loaded.
>
> When they were ordered to bail up White drew his Tranter. A revolver was presented at his head. White jumped at the man that held the revolver, caught him by the wrist and, pointing the Tranter at the fellow's stomach, drew the trigger three times but, to his disgust, without effect. The revolver was not the one White generally carried and he had not noticed that the messenger at the bank had put on the safety guard.
>
> White could not let his man go to unfasten the guard, and so had to wheel him round in front of his mate's revolver and twist him out of the door, where he commenced battering the man's head with the barrel. The bushranger screamed for help so awfully that his two mates came to his assistance and, each holding a revolver at Mr White's head, he dropped the pistol at their feet and said, "There are my arms; take them."
>
> They lowered their revolvers and White drew his second revolver and fired at the first man's stomach within four feet of him. The man fell on his side but got up again. The other shouted, "Fire at the bastard with the double-barrelled gun."
>
> The man with the gun fired direct at White from about six yards away. White, seeing the direction of the gun by the light of the lamp outside the hotel, doubled up and so escaped being hit by the bullet. White fired back but missed and, with only two cartridges left, bolted for the cover of the bush. He hid behind a log near the hotel all night.
>
> Besides Mr and Mrs Booker and family and domes-

tics, there were at the hotel two other lodgers, but none of them were armed. The bushrangers made a thorough search of the premises and inmates, but got only about fifteen pounds. They also took a bottle of brandy and a large plum pudding.

The bushrangers were all masked. They apparently knew White's movements and thought he would be carrying a large amount of money or gold with him. They stayed about half an hour and seemed quite at home in their business. They had no horses with them.

The next victims of the double-barrelled-gun gang were Gold Commissioner Clarke, of Kilkivan, and Dr Mason, of Gympie, who were bailed up on 26 June 1868 on the road from Kilkivan to Boonara, about fifteen miles further to the west. The man shot in the stomach by White had apparently had enough; this time there were only two. Both Clarke and the doctor made a bolt for freedom. The doctor got clear, but Clarke was captured and robbed of his revolver and cash before being released.

Diggers and police scoured the surrounding bush without sighting the bushrangers, but found their deserted camp not far from the scene of the hold-up and took possession of a double-barrelled gun, two revolvers, blankets and some clothing they found there.

There followed two robberies by a different gang—a couple of goldfield thugs who had followed rushes all over eastern Australia and come to Gympie to meet their Nemesis.

On 30 July a miner named Eldridge, riding back to Gympie from the Jimna diggings, was bailed up about midday by two masked men. They made him ride into the scrub at gunpoint, took his horse, saddle and bridle, swag, boots and five shillings and sixpence in cash. Then they tied his hands behind his back, tied his feet together, kicked and beat him about the head, and left him for dead.

About 4 p.m. the same day the two bushrangers bailed up four more diggers, all unarmed, forced them to walk into the scrub, robbed them, and left them tied up to trees. Luckily for them all, one of the diggers had lost the middle finger from one hand and so was able to work that hand free of the ropes that bound him, and free the others. They walked to Imbil Station homestead about five miles away and gave the alarm. A search party found the four victims' swags in the bush, but no sign of the robbers.

The first victim, after spending the night in the bush, managed to crawl, still bound hand and foot, about two miles towards Imbil homestead before being found by station hands.

The average digger and bushman had a strong streak of sympathy for the bushranger and would rarely give him away to the police, but the fact that these two had left their victims tied up in the bush to die put them beyond the pale. Stockmen, station Aborigines and diggers joined the police in the hunt for them, and although the country was ideal for a man on the run, within about three weeks both the robbers had been caught. They were William Trodden, known around Gympie as "Podgy", and Joseph Blake. Both were eventually sentenced to twenty years' jail, the first three years to be served in irons.

Cobb and Co. began to take an interest in the Gympie coach service about the middle of 1868. It was by then a one day run from Gympie to Maryborough in good weather, and the local Cobb and Co. manager, Johnny Myles, after failing to get the mail contract away from La Barte, advertised a passenger service to leave Gympie an hour later and arrive in Maryborough an hour earlier than the existing service. He succeeded in doing it, and soon had the run and the mail contract to himself.

The Cobb and Co. coach left its terminus at the Northumberland Hotel, Gympie, at 6 a.m. and reached Maryborough, if all went well, at 6 p.m. after changing horses at the Nine Mile, Thompson's Flat, Gootchie and Tiaro.

There was still no fixed road, but a number of tracks through the bush, and if the day was fine and the passengers agreeable the trip could be made to take in some beautiful views across country. The main disadvantage of coach travel, apart from the possibility of being bailed up, was that the coach generally left Maryborough heavily laden with stores for stopping places along the way, and on the early part of the trip a passenger could find himself being jostled for position by a bag of flour, a side of bacon, or something similar.

Work had meanwhile been going ahead on a direct road from Gympie to Brisbane. Two road gangs were put on the job, one on the northern and one on the southern section, but progress was slow—partly because of bad weather, partly because of numerous creeks to be bridged and mountains to be traversed, but mainly because every time word came through of a good gold strike at Gympie, gangs shouldered their Government picks and shovels and took the track to try their own luck.

Former timber-getters Low and Grigor, not waiting for the road to be finished, were bringing passengers from Brisbane by ship to a landing on the Maroochy River and putting them ashore there to complete the journey by road.

Brisbane merchant John Francis Buckland, who made the trip in reverse in May 1868, wrote:

> In company with Mr Low I left Gympie at 8 a.m. on Tuesday morning, reaching the Maroochy wharf shortly after sundown; left the wharf at 7.15 p.m., arriving at the mouth of the Maroochy at about 10 p.m.; a walk of from two to three miles over a firm, sandy beach brought us to the steamer *Gneering* at Mooloolah. After some delay we weighed anchor about 1.15 a.m. on Wednesday, making a pleasant run to Brisbane, delivering the mail about noon.

Another route was to take ship from Brisbane to Noosa and go on from there by coach. This was generally a fairly unpleasant trip because the road was very bad, and passengers, already seasick from the sea trip when they started it, often looked more dead than alive by the time they reached Gympie.

The Brisbane-Gympie road was completed towards the end of the year, and the first Cobb and Co. coach was driven through by Hiram Barnes on 12 and 13 November, the 124-mile trip taking two days. On its arrival at the diggings cheering crowds escorted the coach into town and Barnes was carried shoulder high to the Northumberland Hotel for a festive welcome.

K

The Cobb and Co. fare from Gympie to Brisbane was three pounds ten shillings per passenger, children half price. Meals were available at two shillings each. An overnight stop was made at Cobb and Co.'s camp at present-day Woombye, and a bed there cost extra.

Coaches on this run carried six passengers inside and two on the box seat with the driver, who always liked to have a man with him who could help with the brakes while going down the steep pinches. The brakes consisted of a wooden bar across the back wheels with leather pads on the ends to press on the wheel rims, and were operated by a long lever to give plenty of pressure when needed. The four horses were changed at stages about every fifteen miles. Six were used on particularly steep stages.

There were no springs on these early coaches, the body being suspended on leather "braces" or "shanghais" which could stand up to very rough conditions and could be repaired on the road if they did break. They gave the coach a peculiar rocking motion which one passenger described as being like "a baby camel in a hell of a hurry".

It was fairly common to smash a wheel against an unseen boulder or other obstacle, and when this happened the usual thing was to cut a sapling strong enough to support the coach over the bumps but springy enough to cushion them slightly, and lash it under the coach at an angle so it dragged along the ground a few yards behind the coach and supported it like a sort of a skid. In the early days of the service a coach could often be seen skating up Mary Street to the Northumberland Hotel with a sapling in place of one wheel.

Gympie at this time was flanked by two tollgates. At the crossing of Deep Creek, on the Brisbane road, an enterprising townsman named Pengelly had made a bridge by felling two trees across the stream and laying a corduroy decking of logs. Crossing it in a wheeled vehicle was a bone-racking experience. Toll charges were: man on foot, one penny; horse, sixpence; horse and dray, spring cart or cab, one shilling and sixpence; bullock team of ten bullocks, two shillings and sixpence; each additional bullock, threepence; four wheeled vehicle and two horses, two shillings and sixpence; each additional horse, threepence; fat cattle, twopence a dozen; sheep, threepence a dozen.

Though the bridge provided a useful community service, collecting the toll was not a popular way to make a living, and the time came when an irate miner met the demand for toll by throwing his axe at Pengelly's head. Pengelly ducked, received the axe handle across his back, and fell flat on his face. The miner, thinking he had killed him, ran for his life and was never seen in Gympie again.

Pengelly was not seriously hurt, but because of his self-imposed task he was never liked on the field. He was, however, commemorated. Years after the rush was over, a new bridge across Deep Creek was given his name.

On the other side of the town, hotel-keeper J. Heilbronn cleared a road through the scrub at the approaches to the Widgee crossing of the Mary. Gold Commissioner King gave him the right—for a rental of fifty pounds a year—to put up a tollgate. The gate did no more for Heilbronn's popularity than Pengelly's bridge did for his, and tempers became easier in the area when a bridge replaced it.

Bushranging continued. Early in September 1868 the Cobb and Co. coach to Maryborough was held up by three masked men about three miles out of Gympie, and the driver ordered to take it into the bush where the three opened all the mail in search of money and valuables and cleaned out the passengers' pockets.

With the tapering off of the Kilkivan rush towards the end of 1868, that branch of the Bank of New South Wales was closed. The manager, Mr W. S. King, carrying two thousand pounds in half-notes, stopped at Gympie on his way back to Sydney. On the night of 5 January 1869 he was farewelled by the local manager, Mr Buckland, and friends at the Northumberland Hotel, and at 6.30 next morning took the Cobb and Co. mail coach for Brisbane.

There were eight men aboard. In the driver's seat were the driver, Gympie storekeeper Thomas King, and a young digger. Inside on the left was bank manager King, and on his right a digger named Walker and a Gympie business man named Freeston. On the front seat, facing them, were

Gympie's future Town Clerk, Mr J. Kidgell, and a digger who had boarded the coach at the One Mile after being so well farewelled by his mates that he was still hilariously drunk and singing gold-rush songs.

About five miles out from Gympie, after crossing a gully, the coach was slowly ascending the bank when, from somewhere in the bush, there came a sudden command.

"Pull up those leaders!"

As the coachman reined his horses two men stepped from behind ironbark trees on the right-hand side of the road. Each had his face covered with a white cloth in which holes had been cut for mouth and eyes. The more solidly built of the two, the man who had called out, had the driver covered with a revolver. The other, a slightly built man, held a double-barrelled gun at the ready.

The horses had hardly come to a standstill when bank manager King snatched a Tranter revolver from a holster on his belt and, reaching across Walker and Freeston, fired at the man with the revolver. The move startled Walker who jumped up, and in doing so bumped King and spoilt his aim. The shot, nevertheless, took the bushranger in the right shoulder. He staggered back, firing two shots as he did so, ducked behind a tree, and fired again.

The first shot shattered Walker's wrist, the second whistled past King's nose, the third embedded itself in the frame of the coach. The man with the gun had King covered.

"Any more of that and I'll let you have both barrels," he shouted. "Now get out of the coach."

With a movement of the gun he indicated the lad beside the driver. "You," he said, "get to the horses' heads and hold them."

To the driver himself nothing was said after the first call to pull up, and throughout the robbery he was ignored. This, according to police, was the usual professional etiquette among experienced mail robbers.

The wounded bushranger was temporarily concerned with his injury, and in the confusion of getting down from the coach into the roadway, storekeeper King managed to slip his pocket book and a bag containing gold into his top boots, while the bank manager, partly obscured by the men getting out in front of him, slipped his Tranter back into its holster unseen, and pulled his shirt out of the top of his trousers so as to cover it up.

Freeston's main concern was to keep out of trouble. As soon as he was clear of the coach he threw up his arms and called out, "It was not me who fired; I have no revolver; it was him inside"—pointing to King. Nobody took any notice of him, so he said it again and, according to his irate travelling companions, he kept it up right through the robbery.

The drunk from the One Mile was taking the whole thing as a huge joke. After greeting both the masked men affably he decided a song about a bushranger would be appropriate, and embarked on the opening verse of "Bold Jack Donahoe".

The man with the gun ordered the passengers to stand in line behind a log at the edge of the track, take off their coats and waistcoats, put them on the ground and drop their watches and money onto them. The second bushranger stood by, his revolver now held in his left hand, and obviously in considerable pain from his wounded shoulder. It was recalled in his favour later that he made no attempt to molest bank manager King, the man who had shot him.

The passengers, including King, did as they were told and things were going fairly smoothly until the drunk from the One Mile tried to take off his coat, got tangled up in it, and started staggering about the roadway. The man with the gun, apparently suspecting a trick, swung the weapon round to cover him.

King, seizing his chance, grabbed for his revolver.

"Look out," the wounded bushranger yelled. But he was too late.

King got away three shots before the revolver misfired and jammed. One, at least, took effect. With a cry of agony the bushranger dropped his gun and staggered back, both hands grasping his hip.

Still carrying his revolver, King bolted for the cover of a patch of thick scrub about fifty yards off. Two revolver shots followed him but went wide. Then he was out of sight.

The man with the wounded shoulder was half inclined to give chase. "I'll follow that bastard for days if I have to," King heard him say. "I'll shoot him, see if I don't."

But the other had had enough. A red stain was rapidly spreading down the leg of his moleskin trousers. "Get out while we can," he said.

His companion snatched up what money was already on the ground—about twenty-five pounds and a couple of watches—and then both of them, without stopping to search the mails or take any further interest in the passengers, made off into the bush, the man with the gun using it as a crutch and groaning at every step.

Only when they were out of sight did the coach passengers stir themselves to clamber back aboard. Somebody mentioned King, still hidden in the bush. "There's nothing we can do here," the driver told him. "We've got to notify the police." And he whipped up the horses.

From his hiding place King watched the injured bushrangers, still masked, make their way painfully to a spot where two horses were tethered. One helped the other into the saddle and they headed back towards Gympie at a walk.

King watched them until they disappeared from sight over a rise and then, keeping out of sight in the bush, he followed the road on for about two miles to the Seven Mile shanty where he found the coach waiting for him. Storekeeper King borrowed a horse and galloped back to Gympie with news of the robbery, and on hearing it Buckland rode out with a spare horse for King. The two of them returned to Gympie and a hero's welcome at the Northumberland Hotel.

All the police in the district were called out to track down the bushrangers. The day after the robbery two men were arrested on suspicion, remanded for a week, and eventually cleared. Two suspects were pursued and dropped a double-barrelled gun and revolver similar to those used in the robbery, but got away. The *Nashville Times* reported:

> On Thursday morning 7 January Detective Smyth and a constable searching near the Maryborough Road, approached a house where they could see a saddled horse. A man was watching them approach. Next thing, the wanted man emerged, sprang onto the saddled horse, and galloped away. Smyth, drawing his revolver, shot himself through the left hand. After that he tried to shoot at the man but the revolver would not go off.

The constable fired two or three shots but missed the man, who was about 200 yards off. The police, being on foot, could not follow. They arrested the householder on a charge of harbouring a bushranger.

This man later established his innocence and was released. When the next arrest was made, neither police, press nor public waited for the law to take its course before deciding on his guilt. The *Nashville Times* reported:

> January 14: Excitement was caused by the arrival in town of a mounted trooper with the news that one of the bushrangers who stuck up the Brisbane coach had been arrested. About midday yesterday Sergeant Calloppy discovered a man camped in the scrub near the racecourse grandstand. The man, who was wounded in the left hip, said he was hurt when his gun accidentally went off. He was arrested and brought to the lock-up in a dray.
>
> January 22: The man Bond, charged with bushranging, has been remanded for the evidence of Walker who was wounded and is now in hospital. The case is clear as to Bond's identification. Two witnesses have sworn to him, and he can give no satisfactory account of his own wounds.

When William Bond eventually came before the court on a charge of robbery, evidence was given that passengers from the coach had identified him by his voice in a line-up of twenty. He protested his innocence but was convicted and sentenced to jail for twenty years, with two whippings in the first six months of his sentence.

What happened to the other bushranger is a much longer story.

Chapter Sixteen:
Bushranger's Conscience

Though the police were not sure of it at the time, the bushranger behind the Gympie coach robberies and the man who was wounded by bank manager King's first shot was a wild young colonial named George Palmer.

In his early twenties, sturdy, handsome, and related to the Campbells of Campbell's Wharf, Sydney, he had opportunities above the average when he arrived at Rockhampton some time after the Canoona rush and took a job managing an out-station for W. H. Holt. Soon he was known as a good bushman, a splendid horseman, and a man who was game to try anything.

Finding station life too tame for his taste, he drifed to Rockhampton, where he quickly got a reputation as a reckless hotel brawler. He married an extremely attractive seventeen-year-old girl of the town, nearly killed several men who felt they had long-standing claims to her favours, and, early in 1868, when the rush to Gympie was nearing its peak, stole a fine grey horse named Bobby from Morinish Station and rode south.

His wife followed him as far as Maryborough and there appeared to live reasonably well, with little to worry about but her husband's inordinate jealousy.

There is nothing to show that Palmer ever dug for gold at Gympie, though he often seemed well supplied with it. Police wanted him for horse stealing and suspected him of several robberies. In the district it was soon known that he was the man behind the early coach robberies, but, like many other bushrangers who kept to the rules, he was not unpopular and shared many a billy of tea with bushmen, carriers and others who knew very well who he was.

One of his hiding places was supposed to be a cave in the sandstone near Eel Creek, which joined the Mary River just downstream from the diggings.

Zachariah Skyring used to recall that when his family first came to Gympie there was such a demand for sawn timber and split palings that his father could not buy any, so he took young Zac with him in the dray and set up a camp in the bush near Pie Creek, a tributary of Eel Creek, to split his own.

One morning they were boiling the billy when a black-bearded horseman rode up. Skyring senior invited him to a drink of tea, which he accepted. As the man squatted at the fire and leaned forward to take the pannikin of tea, he revealed a pistol holster on each hip.

In the act of lifting the pannikin to his lips, he paused and seemed to listen intently. Then, without a word, he took a flying jump straight for his saddle. As he landed in it his stolen horse, Bobby, as though knowing what to do, started to move off, and by the time the rider was settled was at full gallop, tearing through the scrub as fast as he could go.

Man and horse came into sight again as they crossed an open patch heading for the creek.

From somewhere in the bush there came shouts of, "Halt; halt in the name of the law," followed by about half a dozen shots.

But the stranger rode straight at the steep bank of Pie Creek, put his horse down a jump of about ten feet, rode him up the opposite side, and was away.

Then the police troopers came in sight, still firing, but too far behind to have any chance of hitting their target. When they came to the creek none of them would put his horse to the jump. They rode back to Skyring's camp to make inquiries and told him the man who had got away was Palmer the bushranger.

On another occasion two pit sawyers at Eel Creek were having a meal in their hut on the flat when the black-bearded man arrived and was invited in for a drink of tea. He came in but sat facing the door of the hut so he could see what was going on outside.

While he was there Inspector S. J. Lloyd, of Gympie, and Constable Tommy King came riding across the flat. The man dived from the hut, sprang into the saddle and was away in seconds. The police saw him and gave chase along the track leading to the creek. There was a big tree lying across it but the stranger took it at a jump. The police horses refused to take it and by the time the pursuers had gone round it through the thick scrub their quarry was gone.

Storekeeper Southerden, who had once been doubtful of buying Nash's first gold, was by now taking large amounts of gold from diggers in payment for supplies, and he regularly forwarded it to Maryborough by any traveller willing to oblige.

Digger Edward Poulton, who rode regularly to Maryborough to see his family there, left on one of his trips carrying forty ounces of Southerden's gold. A short distance out of town he caught up with a bullock team, rode alongside for a while and was talking to the driver when another horseman overtook them. He passed and then deliberately turned his horse across in front of the bullocks and came back on the other side so as to bar Poulton's way. Poulton rode round the man without speaking to him and as he and the bullocks continued on, the man sat on his horse in the middle of the track looking after them as though not sure what to do.

"Know who that is?" asked the bullocky.

"No."

"That's Palmer the bushranger. He must think you're carrying gold. If you hadn't been with me, you'd most likely have been stuck up."

Never again did Poulton agree to carry anybody's gold to town for them.

After Bond's arrest for the Gympie coach hold-up, Palmer headed back for Rockhampton. He arrived early in February and there took up with some of his old mates. Among them was Alexander Archibald, a horsebreaker and dealer who had lately taken over the Lion Creek Hotel, just out of Rockhampton on the track to the north-western goldfields. Archibald had won a good many races with a pony of his called Quart Pot, and a man named Charlie Taylor trained for him.

Archibald was generally regarded as a likeable character, though not above a shady deal that showed promise. He had a wife and two children, all three of whom were good but reckless riders. Taylor was good with horses but had had some trouble with the police in the south and was regarded as a shifty type not to be trusted any further than necessary. These two and Palmer were all men of about twenty-five at the time.

Palmer made his camp further out along the track at a spot called the Agricultural Reserve, a few miles up the river from Rockhampton. There were others camped there and they knew Palmer was wanted for robbery under arms, but all were more likely to warn than betray him.

The only narrow escape he had was soon after his arrival. Horses and vehicles were taken across the Fitzroy River at that time by a large punt propelled by long sweeps. One evening Palmer led his horse onto the punt and it was about to push off when two mounted policemen rode up and asked the lessee, Frank Humphreys, if he had seen or heard anything of Palmer, whom they wanted for horse stealing.

Humphreys said no, and they said they were coming over on the punt. As they rode their horses aboard, Humphreys, who knew Palmer well, whispered to the bushranger to take one of the sweeps and row for his life. Palmer did so and made such a good job of it that the police ignored him and rode off as soon as the punt grounded. Palmer thanked Humphreys and followed them at a safe distance.

Many different stories were told of what followed, as each of the four men involved tried to shift the blame from himself onto the others. The most accurate is probably that given by Palmer himself.

According to Palmer he called at the Lion Creek Hotel the evening he arrived back in Rockhampton, and Archibald told him there was a man named Williams waiting for a chance to stick up a gold buyer named Patrick Halligan. Archibald wanted to know if Palmer would be in it. Palmer said no. He knew Halligan, who had been a wild man himself in his day, and now, aged thirty-one, with a wife and four children, and owner of the Golden Age Hotel in Rockhampton, was more than ever a man to be reckoned with. He always carried a revolver and regularly boasted that anybody who ever tried to rob him would get a bullet through the skull.

Archibald persisted and arranged a meeting between Palmer and Williams three days later and, after a good deal of argument, Palmer agreed to join in the proposed robbery. In doing so, though he apparently did not realize it, he got himself into more desperate company than even he was accustomed to.

John Williams, a native of New Zealand, was better known around Rockhampton as Old Jack. He was about fifty, bearded, a rough man, but a thinker. He was well known as a con man, and was regularly to be seen around hotels, horse sales, or any other places where money was changing hands. To the local people he was something of a mystery, and some said he was Thunderbolt the bushranger. This could not have been correct, but he was that sort of man.

After a plan had been decided on Old Jack moved out to the Agricultural Reserve with Palmer to wait for word from Archibald. The first time Halligan came through they missed him because Archibald had not warned them when to expect him. On another occasion bank manager T. S. Hall, who also bought gold on the fields, came through and Old Jack wanted to stick him up, but Palmer refused. With Palmer fairly lukewarm about the whole thing, Old Jack went to town to do some scouting on his own account, and walked right into the chance he had been waiting for.

Halligan, who never made any secret of his movements, seemed to go out of his way to provide it. On Saturday, 24 April 1869, he told everybody in the bar of the Golden Age that he was going out next day to buy gold at the Morinish goldfield, about thirty miles to the north-west of Rockhampton and that he expected to do well out of the trip.

Old Jack, who was in the bar at the time, and no doubt hoping to clear himself of suspicion of implication in future events, gave him a warning:

"You'd better hold your tongue, Halligan, or you'll get stuck up one day and probably get a bullet through your head."

Halligan laughed and slapped the revolver he always carried in a holster on his belt.

"Nobody will ever get any gold from me as long as I can pull a trigger," he said.

Next morning Halligan called in at Archibald's Lion Creek Hotel for a drink on his way out to Morinish. He told Archibald where he was going and that he would be riding back that night. Old Jack was lounging on the veranda at the time, and after Halligan had gone he took his own time in riding out to join Palmer at the reserve.

Halligan rode on to Morinish where he bought a sixty-seven-ounce cake of retorted gold—much less than he had expected—from the Alliance Company, and set off on the ride back to Rockhampton. He was seen about half an hour before sunset at a creek crossing between Ridgelands goldfield and the Agricultural Reserve, and shortly after dark must have reached a patch of brigalow scrub close to where the track passed the reserve. Palmer and Old Jack Williams were waiting there for him.

According to Palmer's story there were three or four tracks running nearly parallel through the brigalow. He and Old Jack were hidden in the bush beside one of them and, just about dusk, they saw a man riding towards them along the next track at full gallop. Palmer could not recognize him because his eyes were bad with an attack of the blight, but Old Jack had no doubts.

"That's Halligan," he said.

As the rider came opposite where they were hidden they spurred their horses and galloped along their own track parallel to him for about two hundred yards. The man saw them then and called out. Palmer did not recognize the voice.

"I don't reckon that's Halligan," he said.

Old Jack would not have it. "I'll take my bloody oath it's him. Keep on going."

They kept on and as soon as they came to a spot where the brigalow thinned they cut across through it so they came out about twenty or thirty yards in front of the galloping rider.

Old Jack called to Palmer to pull up, but as Palmer reined his horse Old Jack's animal ran into him and his horse bolted. By the time he pulled up he heard Old Jack behind him shout.

"Bail up, Halligan!"

Palmer wheeled his horse and galloped back. Jack was holding Halligan's bridle on the near side and Halligan was fighting to free it. With a sudden wrench he broke away and sent Jack reeling. Palmer swung in on the opposite side of Halligan's horse, tugged out his revolver and aimed at Halligan's head.

"Stand!" he shouted.

Halligan turned and found himself looking down the barrel.

"I know you, Palmer," he yelled, then, as though regretting what he had said, "Don't shoot and I'll say nothing."

Without waiting for an answer he clapped his heels to his horse's flanks and leaned low on the animal's neck. The horse leapt forward. Palmer spurred beside him, dropped his own reins and grabbed Halligan's bridle with his left hand. With the other he jabbed the revolver again and again in Halligan's ribs, but made no attempt to fire. Both horses had swerved from the track and were crashing through the brigalow.

"Stand and I'll put the gun away," Palmer yelled. "I want to talk to you."

Halligan only urged his horse on.

"I'll not give you the gold, Palmer," he yelled. "You'll never get the gold."

He lashed out savagely with his riding whip and caught Palmer a stinging blow across the face, almost blinding him temporarily, and before the other could recover, brought the whip down on the hand that held his bridle.

Palmer lost his grip and reeled in his saddle as the horses separated. Halligan dropped the whip and reached for his own revolver.

From further back in the bush came a warning shout from Old Jack.

"Look out, Palmer; he's pulling his gun. He'll shoot you."

Halligan's revolver cracked and Palmer heard the bullet whistle past his head.

There was another shot then, and Halligan, hit, reeled in his saddle and dropped his revolver. But he still rode on, yelling at the top of his voice, "I know you, Palmer; it's Palmer, Palmer, Palmer."

At last the two galloping after him saw him roll sideways and fall from the saddle. As they came up with him Palmer jumped off his horse and lifted him over against a tree.

"I'll get a doctor," he said.

Old Jack stared at him in amazement.

"Don't be a bloody fool. He's going to die anyway. Get the gold."

Halligan was still conscious. "I know you, Palmer," he gasped. "I know you, Palmer; you'll hang for this."

They gagged him to keep him quiet. Then they took the cake of gold, went through his pockets and took some banknotes, and took a gold ring from his finger. They picked up Halligan's revolver from where he had dropped it and rode away, leaving Halligan on the ground, still gagged and bleeding to death.

Palmer and Old Jack rode back to the Lion Creek Hotel where they found Taylor and told him what had happened. Archibald arrived back from town about 10 p.m. and they told him too and showed him the gold. Archibald, white faced and terrified, stepped back from the gold in horror.

Palmer, who was himself more wrought up than anybody had ever seen him before, turned on him in snarling fury, "Go on, take hold of the bloody doughboy, it won't bloody bite you. You were game enough when you were telling us to do it."

They all argued about what to do next. Then Palmer and Old Jack took a bag and rode back to where they had left Halligan. He was dead. It was about midnight by then. They caught Halligan's horse, lifted the body across the saddle and tied it there, and took it across country to the river There they crammed it into the bag they had brought and weighed it with bricks from the chimney of an old, deserted house. They put it back on the saddle and led the horse down through the reeds that lined the bank of the river and dumped the body and saddle into the water. It was hardly deep enough to cover them and Palmer stripped and waded in and dragged them into deeper water.

They camped for the night and next day led Halligan's horse away into a lonely part of the bush, shot him and cut out the brands, which they buried.

The two of them remained in hiding in the bush until 5 May when Archibald and Taylor came out and met them to divide the spoils. Taylor brought a spring balance and a tomahawk, and Archibald brought some brandy. They drank the brandy and chopped the cake of gold into four parts. A piece weighing about twenty-five ounces went to Palmer, and one of twenty-four ounces to Old Jack. Archibald got twelve ounces and Taylor got six. Twelve pounds in notes taken from Halligan's pocket was divided between Palmer and Old Jack, and they tossed for Halligan's gold ring, which was won by Old Jack.

Next day Palmer left for his old haunts at Gympie, leading a stolen pack-horse on which was a saddle belonging to Archibald.

When there was still no sign of Halligan on Monday—the day after he should have been back in Rockhampton—Gold Commissioner John Jardine organized a search to go over his tracks. Eventually Halligan's hat, whip, and a piece of his coat were found at the scene of the robbery.

Headquarters of the search party, ironicially enough, was the Lion Creek Hotel, where Archibald sweated it out in terror of discovery of his part in the robbery, but said nothing. The party found the tracks of Halligan's horse and two other horses, but after that a week went by with nothing further discovered.

The Government offered a hundred pounds reward for information. Still nothing came to light, and the reward was raised to three hundred pounds; then citizens opened a fund that soon collected more than four hundred pounds.

Halligan's body was found on 7 May by a boat party led by Frank Humphreys, the ferryman who had helped Palmer escape the police about three months earlier.

The police had regarded Palmer as a suspect from the start, but they had nothing to go on except that he had been back in the district at the time of the robbery and had since left. The first clue meant little at the time. Before Palmer reached Gympie his stolen pack-horse, still wearing Archibald's saddle, broke away from him and was found by the police. The horse and the saddle established a link between Palmer and Archibald.

Then Sub-Inspector Elliott received a letter from a digger on Ridgelands goldfield, on the north-western fringe of the Agricultural Reserve, saying he could tell the police something but was afraid to come to town to talk to them in case some of Palmer's friends should see him and kill him for giving Palmer away. Elliott went out to the man's camp and learnt that the digger had seen Palmer waiting, apparently to waylay Halligan, on the night of the murder.

The police, unable to lay hands on Palmer, and acting on the saddle clue, arrested Archibald, accused him of helping Palmer to escape, and soon had him in such a state of panic that he told them the whole story in the hope of saving his own neck. Old Jack and Taylor were arrested in Rockhampton, and Inspector Uhr and his Native Police were sent south to help in an all-out hunt for Palmer.

Palmer had moved fast, but had left a fairly clear trail leading back to his old haunts in the Gympie scrub. At Calliope he had displayed a blood-stained revolver and said he had been stuck up by bushrangers and had to shoot a man. As there was no reason for him to make up such a story, it seems likely that the death of Halligan was preying on his mind so that although he did not admit to having had any part in it, he had to talk about having shot somebody.

Later on, at the Burrum River, on 13 May he told a barmaid in a shanty that he was a gold buyer, showed her his lump of retorted gold, and said he had lost the rest when his pack-horse broke away near Rockhampton. He asked her if she would like to have a piece of the gold to have a ring made. She said she would, and Palmer got a tomahawk, chopped off a piece weighing about an ounce, and gave it to her. When the police came asking questions the girl showed them the gold, which they at once saw was retorted gold like that stolen from Halligan.

By the time Palmer reached his cave on Eel Creek, the police were on his track and there was a price totalling nearly eight hundred pounds on his head. He himself felt he had a good chance of remaining at liberty. He knew the Gympie scrubs well, and was confident he still had friends in the district. It was soon obvious that he did. But then Palmer found that he had a totally different kind of problem to deal with.

Throughout his career as a bushranger Palmer had never given any indications of being a killer. His reluctance to shoot to kill had probably spared the life of White during the affray at Currie Hotel, and during the mail-coach robbery, though wounded by King, he had made no attempt to harm King during the brief period when he had the bank manager at his mercy.

Even on Old Jack's story, Palmer had shown a reluctance to shoot, and when Halligan fell, Palmer's first thought was to get a doctor.

Halligan's death preyed more and more on his mind. He brooded and began to wonder what use it was to keep running. Adding to his worries was the fact that his young wife, apparently in the hope of being able to help him, had moved from Maryborough to Gympie. Her earlier reputation as a girl of the town had preceded her and diggers crowded round to press their attentions. Friends reported this to the always jealous Palmer who, holed up as he was in a cave on Eel Creek, could do nothing about it.

Then a mate named Bluey came and told him that a Detective Hanley, one of the men sent to Gympie to hunt him down, was paying persistent attention to Mrs Palmer, and was apparently meeting with some success.

"I'll kill the bastard," said Palmer frantically. "I don't care if every policeman in Queensland is there, I'm going to town tonight to kill him."

Bluey managed to quieten him down at last. He said he did not think Mrs Palmer knew Hanley was a detective because Hanley had tried to make a secret of it. He said it would be best if Palmer stayed out of it. "You've got mates of your own in town, George," said Bluey. "We'll deal with Hanley our way."

Things came to a head a few nights later at a dance-hall. The story was told months later to Mining Warden W. R. O. Hill by none other than the discoverer of the Palmer River goldfield and veteran of a dozen rushes, James Venture Mulligan. Mulligan recalled:

> Just before leaving Gympie for the north Dr Jack Hamilton, who was a gold digger then, and myself entered the passage at Billy Flynn's Hotel on our way to the dance room. As we pushed along a man in front called Bluey accused a man beside him of burning his finger with a cigar. The man replied that if he had he was very sorry. This failed to satisfy Bluey who insisted it was done purposely.
>
> Hamilton told him he should accept the apology as it was evidently unintentional. Bluey was, however, determined on blood, and insisted on fighting. The man replied that he was a stranger and could not depend on fair play. Hamilton then said, "I'll second you and you will get fair play." A ring was made in the middle of the dance room, the women standing on the chairs and forms, the inner circle of men squatting on the floor.
>
> At the call of "Time" Bluey sprang from the knee of his second, Long Bill, and the stranger from Hamilton's. The stranger dropped Bluey and directly he fell Long Bill rushed at him.
>
> Hamilton cried "Fair play" and sprang in front of the stranger. Long Bill let go his left at Hamilton who let it shave past his cheek and landed his left with such force that the first part of Long Bill's anatomy to touch the floor was the back of his head. Bill was then pulled to a corner and before he regained his senses, the dancing was again in full swing.
>
> Years later, on the Palmer, Hamilton told me the sequel. On his way home the stranger overtook him and said: "You saved me from being mobbed this evening and I think I should tell you who I am as I am sure you will not divulge the name. I am Detective Hanley. That young woman I danced with so frequently is Mrs Palmer. Her husband is wanted for murdering Halligan. Those fellows are jealous of her preference

for me, and that row was planned tonight as an excuse to mob me. I am merely making love to her professionally to get news of her husband. She says he visited her last night and threatened to shoot me as he is jealous of me too."

The result of the fight and Hanley's continued attentions to his wife further depressed Palmer, who spent more and more of his time brooding in his cave.

His remorse over the shooting of Halligan apparently spread to the way he had treated his wife, who had defended her own actions by accusing him of neglecting her. The time came when he could stand it no longer.

One of Gympie's prominent solicitors at the time was J. W. (Wicky) Stable, who had built himself a comfortable home about two miles down-river from the town, not far from where Eel Creek flowed into it. Early one morning Stable answered a knock on the door to find himself faced by a man dripping wet from having swum the river.

"Are you Mr Stable, a solicitor?" the man asked.

"Yes I am."

"I wish to see you on a confidential matter."

"Come inside."

The man took the offered chair. His face was drawn, his eyes sunken, his hair and beard unkempt, and his hands trembled. He was making an effort to speak calmly.

"I'm Palmer the bushranger," he said.

After a strong nip of brandy Palmer put his plan to Stable. He was tired of running, he said. If he undertook to give himself up, would Stable be prepared to bring the police to his hiding place, claim the reward for himself, and use part of it to defend him and give the rest to Mrs Palmer?

Stable asked for time to consider the suggestion. He checked with the police to make sure that the offered reward was for Palmer's apprehension, not for his conviction, and on being told this was correct, agreed to Palmer's plan. He informed the local man, Inspector Lloyd.

As the inspector and his men approached the spot described, they saw Palmer sitting on a rock. Lloyd covered him with two pistols and called on him to surrender. Wearily Palmer got to his feet.

"You can put those guns away," he said. "If I'd wanted to, I could have shot you before you even saw me."

What really happened to the reward money remains a mystery. Some believed it was paid to solicitor Stable and that he carried out the terms of his agreement with Palmer; others said it was paid to a policeman. Mrs Palmer went to Rockhampton for her husband's trial, and soon after left for Sydney.

Palmer and Old Jack Williams were convicted of Halligan's murder and sentenced to death. Throughout their trials each claimed the other had fired the fatal shot. Archibald was convicted of inciting and aiding them and also sentenced to death. All three were hanged.

No charge was proceeded with against Taylor and he was released, after having given evidence against the others.

A few minutes before he was hanged Palmer cleared his conscience of the weight which had finally broken his reckless spirit. In a confession signed before Police Magistrate Wiseman he stated: "I shot Halligan."

There was one other man on whose conscience some responsibility for Halligan's death rested for the remainder of his life. The ferryman, Frank Humphreys, who had helped Palmer elude the police troopers on his ferry, and had later dragged Halligan's bloated body from the water, recalled both incidents years later.

"If I'd only turned Palmer over to the police that night, perhaps neither Halligan, nor Palmer, nor Old Jack, nor Archibald would have died," he said.

Chapter Seventeen:
Bringing in the Big Cake

The five years that followed the first crushing of ore in April 1868 saw the development and exploitation of the rich, shallow reefs of Gympie's brown slate country. By April 1869 Gympie mines were keeping a hundred stampers going. By the end of the year ore was being brought from a depth of 150 feet or more. Gold sent out by escort in 1869 totalled 70,582 ounces; in 1870, 44,159 ounces; in 1871, 44,712 ounces; and in 1872, 48,964 ounces.

Writer and traveller Anthony Trollope, who visited Gympie in August 1871, found it "a marvellous place, very interesting, though very ugly", and went on:

> The main street contains stores, banks, public houses, a place of worship or two, and a few eating houses. They are framed of wood, one storey high, generally built in the first place as sheds with gable end to the street, onto which, for the sake of importance, a rickety wooden façade has been attached. The houses of the miners, which are seldom more than huts, are scattered over the surrounding little hills, here and there, as the convenience of the men in regard to the different mining places has prompted the builders. All around are to be seen the holes and shallow excavations made by the original diggers, and scattered among them the bigger heaps which have been made by the sinking of the deep shafts.
>
> Where a mine is being worked there is a rough wooden windlass over it, and at a short distance the circular track of the unfortunate horse who, by his rotary motion pulls the buckets up with the quartz, and lets them down with the miners. Throughout all, there stand the stunted stumps of decapitated trees, giving the

Bringing in the Big Cake | 155

place a look of almost unearthly desolation. At a distance beyond the mine shafts are to be seen the great forests which stretch away on every side.

There were, perhaps, fifty or sixty reefing claims in which mining was actually in progress, but I did not hear of above ten in which gold was being found to give more than average wages, and I heard of many from which no gold was forthcoming. This claim had been abandoned, that other was about worked out, a third had been a mere flash in the pan, at a fourth they had not got deep enough and did not know that they ever would, at a fifth the gold would not pay the expenses.

The gold coming from the good mines was, however, enough to give the town, which had settled down to a fairly stable population of something less than five thousand, a general feeling of hope and prosperity. The mines were mainly in the hands of small companies whose shareholders and directors were themselves working miners. Before long comfortable homes began to dot the slopes of Gympie's seven hills.

Exciting stories came out of Gympie in those days—stories like the epic of the "Big Cake", a great slab of retorted gold from the Monkland reef, weighing 5,972 ounces, worth, on values of the time, about twenty thousand pounds, and on present day values more than a quarter of a million dollars.

The Monkland reef, then on the southern fringe of the field, was opened up soon after the Caledonian. The Monkland P.C. itself was not spectacular, but other claims on its line soon made the name famous. Best known of them all was the mighty claim that came to be known as No. 7 and 8 South Monkland—producer of the Big Cake. The No. 7 claim, 160 feet on the line of reef, was originally applied for on 30 May 1868 by L. North, Stephen Ryan, Robert Jones, Michael Jackson and Frederick Braddell; and No. 8, 240 feet, on 13 June 1868 by James Glasgow, John Smolden, Jacob Pearen, Daniel Smolden, Joseph Pearen and William Bath. Glasgow's share was soon transferred to George Beer and Bath's to John Pearen.

Then No. 7 and No. 8 South Monkland amalgamated, giving four hundred feet on the line of reef and making for more economical and efficient working. Rewards came quickly. On 15 April 1871, 212 tons of ore yielded 3,984 ounces of gold.

Developmental work went ahead, and on 20 July 1872 a crushing of 439 tons yielded 4,028 ounces of gold. It was the largest return, up to that time, from a single crushing, but the next one surpassed it. On 20 November 1872, 739 tons of ore yielded 5,972 ounces of gold and a further batch of ore from the same crushing, an additional 112 ounces, making a total yield of 6,084 ounces. It represented fourteen weeks' mining by the sixteen men who were working the combined claims at that time, and for more than ten years remained Gympie's record crushing.

The 5,972 ounces of gold came from the retort in a giant yellow cake too heavy for a man to lift. It was taken to the bank in a dray. Miners coming to gaze at it in awe christened it the "Big Cake".

The owners' problem was to get it to market, local gold prices still being below what most miners expected. The escort fee from Gympie to Brisbane was sixpence an ounce, so the price for taking it by escort would have run out at £149, not enough to make a large hole in £20,000 perhaps, but, in the owners' opinion, too much to pay out to an escort service whose high charges had been a source of irritation ever since it began. In any case, there was something about the look of the great slab of gold that made one reluctant to trust it to others, and the owners also felt like doing something spectacular to celebrate.

Conferences were held and finally a coach, driver, and relay teams were hired for forty pounds to take the Big Cake to Brisbane, the owners themselves to ride escort.

So the cake was put in an iron retort pot and the whole packed with earth into a large, strong packing case. This, with a good deal of sweat and manoeuvring, was eventually loaded onto the floor of the coach. The final celebration was held at the Northumberland Hotel with owners taking turns to guard the case. On the morning of 24 November 1872, the four-horse team was harnessed and the coach moved off to the blast of brass trumpets, with two shareholders riding inside armed with pistols to keep an eye on the case, another on the driver's box armed with a carbine, and eight more, mounted on good horses and all carrying carbines, riding before and behind. Rarely had there been such a rich haul for a party of bushrangers, but it was guarded by probably the strongest posse in the colony.

The coach trip to Brisbane was two days of hard driving over about 124 miles of difficult country. Often there was no road at all and the coach was driven at random through the bush, taking whichever of about half a dozen tracks appeared best at the time.

There were fallen trees to be chopped up and dragged clear, there were running creeks with approaches of deep mud and bottoms of boulders, there were slopes to be skirted where only the barest pretence of a cutting had been made and the coach was in constant danger of rolling over down the mountainside, and there were pinches so steep that passengers normally had to get out and carry their luggage while the horses laboured to pull up the coach. It was impossible to drag out a packing case weighing about four hundredweight and carry it up every steep pinch, so the escort had to get off their horses and heave at the wheels.

At the end of the first day's travel they reached Cobb and Co.'s overnight stopping place at Woombye, hot, dry and hungry, glad to hand over their horses to stable hands and troop inside for a drink before dinner. Several pots had been drained before one of them halted his tankard half-way to his lips, slammed it down on the table, and bolted for the door.

"The gold," he yelled. "We forgot all about the gold."

A stampede to the unguarded coach revealed the packing case with the Big Cake in it still safe inside. It would, in fact, have taken a team of men to steal it, but two men were left to watch it, and two remained on guard in relays throughout the night while the rest slept round the coach.

On arrival in Brisbane the Big Cake was lodged in the Queensland National Bank for safe keeping. Until then its true weight was not known, and when it came to the weighing it was found there were no gold scales in the town that could handle such a mass. It was weighed on avoirdupois scales at last, and tipped the balance at three hundredweight, two quarters, nine and a half pounds.

Exhibition of the Big Cake at the Brisbane School of Arts caused a sensation. Such a huge mass of gold had never been seen in any of the colonies, its closest rival, the Krohmann Cake which had been exhibited in Sydney in May 1872, having weighed no more than 5,612 ounces. Crowds flocked to see it at sixpence a head admission, proceeds going to the Brisbane and Gympie hospitals.

Its journey from Gympie to Brisbane conferred one other benefit on the mining community. Gold escort fees were quickly cut from sixpence an ounce to fourpence halfpenny.

No. 7 and 8 South Monkland remained one of Gympie's big gold producers for years. Notable among its yields were: 9,398 ounces of gold from 3,210 tons of ore in 1873; 5,001 ounces from 2,305 tons in 1874; 2,149 ounces from 1,243 tons in 1875; and 6,150 ounces from 3,109 tons in 1876.

It brought several fortunes to its shareholders, among them Jacob Pearen who used some of his gold to build a unique monument that survived the goldrush by nearly a hundred years.

Jacob was a Cornish miner who came to Australia about 1860 and, although he remained a miner for most of his working life, he was fascinated by ships and the sea. It was this that led him, in 1882, to build "The Lighthouse" on Redcliffe Peninsua, looking across Moreton Bay and covering the entrance to the Brisbane River.

Victoria House—to call it by its correct name—was a wooden house, four storeys high. The third storey consisted of a garret, and perched on top of it was Pearen's own cabin with a balcony from which he could look out to sea and a light which he burned continually as a guide to mariners. It was not an official light, but it became so well known that sailors used to watch for it, and whenever a ship anchored off the point a boat or two would soon row ashore.

In his later years, when his wife had died and his daughters had married, Jacob became an aloof, lonely man, regarded with awe by local Aborigines but with affection by sailors who always found a warm welcome in his home. Whenever there were ships in the bay the lighthouse would be ablaze with light from all floors and resounding with chanties and stamping feet as guests found their land legs after long days at sea.

Above the second-floor balcony a strong beam and pulley projected from the roof, and brewery wagons drew up beneath it while their cargoes of casks were slung up onto the upper balcony. In the first light of the dawn that followed a party, it was said, unconscious sailors could be seen being lowered by block and tackle to be doused with water at the tankstand before being returned to their ships. Stairways inside the house were steep and narrow like ship's companionways, and not for men unsteady on their feet.

Old Jacob remained in his lighthouse until he died in 1916, aged seventy-eight, and after that it was taken over by his nephew who lacked the old man's love of the sea. The old hospitality faded away and the lamp in the lookout was not lit.

After passing through a number of hands, the house was at last left vacant and began to fall to pieces. In 1967 it was put up for auction and passed in at $7,400. An airline pilot bought it in February 1968 with the intention of demolishing it to build on the site. Local historians took a belated interest and asked the Redcliffe council to take it over for a museum. The argument was still going on when the old house was destroyed by fire in June 1968.

Though a substantial slab courthouse had by this time been provided, mining law in Gympie remained on a fairly shaky basis throughout the whole of the first boom period. Mines continued to operate under the New South Wales law, which often had little application to local problems. In an attempt to help the overburdened Gold Commissioner, a poll was held in November 1869 to elect a nine-member Gympie local mining court consisting of practical miners.

One of the leading candidates was Dr Theodore Edgar Dickson Byrne, a bouncing, colourful character better known as the "Jumping Doctor". He was against the Government on things like fees, rates and taxes, and rarely silent on anything that could be argued about. He regarded himself as the "diggers' friend" and a lot of diggers regarded him as a nuisance.

Bill O'Regan, one of his rivals for the court position, advertised, "Vote for the Jumping Doctor if you want to be bled". J. D. Collis, another candidate, advertised, "Byrne, Bounce and Bunkum; O'Regan, Rot and Remorse; Collis, Confidence and Commonsense".

O'Regan attracted his audiences by decking himself out like a maypole in bright ribbons, Collis cantered around the town on horseback balanced on his head. Byrne, more restrained than anybody had ever seen him before, confined himself to oratory.

Byrne won and quickly bounced back to his old form, telling Gold Commissioner King at the declaration of the poll that the Commissioner was quite a nice fellow, but some of his decisions on mining disputes were quite illegal.

Among others elected to the court were the legendary Doctor Jack Hamilton and his mate, James Venture Mulligan. Registrar of the court was Robert Critchley, who had arrived from Victoria in October 1867 bringing with him a copy of the Ballarat mining regulations. The book was the only authority of its kind on the field, and time and again, until the Goldfields Act of 1874 became law, Bob and his book saved Gympie from chaos.

The town's first solicitor, Mr (later Sir) Horace Tozer, soon became an authority on mining law and won considerable popularity among the miners by solving problems for them. He went on to become alderman, M.L.A., Colonial Secretary and, in 1898, Queensland's Agent General in London.

The showplace of the goldfield in the 1870s was the New Zealand P.C.—or Maori mine, as it came to be called—where the quartz always had a good showing of gold. When the Governor of Queensland, the Marquis of Normanby, visited the field at the end of April 1873, great preparations were made to entertain him at an underground banquet in the Maori mine.

The walls of the mine were swabbed down so the crystal and gold in the reef would glisten in the light of hundreds of candles mounted in clay. A huge banquet table was built at the 275-foot level, chairs were provided, and the directors' wives were lowered in the buckets by horse-driven whim to

Bringing in the Big Cake | 161

lay out the linen, silverware and crystal, and masses of freshly cut flowers. An ample cellar was laid in, ice buckets prepared for the champagne, and arrangements made to lower each course of the banquet, which was being prepared by a roster of cooks.

But things did not go altogether according to plan. The Governor arrived, was received with speeches of welcome, and was taken in a buggy on a triumphal procession through the town to the One Mile. Roads were still very bad, and on the way back part of the buggy frame broke and a sharp piece of wood jabbed the horse, which bolted. Unable to hold him, the driver steered into an embankment. The vehicle overturned and the occupants were thrown to the roadway. The driver was severely injured, and the Governor received a sprained ankle and such a severe shaking up that he had to rest for several hours before returning to his hotel where he remained for the rest of the day.

The underground banquet went ahead with the Attorney General (Hon. J. Bramston) deputizing for the Governor and, according to a contemporary report, "several toasts were honoured by a gathering of thirty-five, including a number of women".

Some of those who waited above later recalled, however, that as time went on and nobody appeared at the pit-head, some anxiety began to be felt about what was going on down below, and a messenger was sent down the mine to inquire discreetly if everything was all right. More time passed, but the messenger did not reappear. A conference was called and a decision had just been reached to send another man down when the signal came from below to haul away. The first of the returning revellers reached the surface singing lustily and announced that they would be "down there yet if the brandy hadn't given out", and that the others were now ready to be brought up.

In spite of the unfortunate absence of the Governor, the banquet was such a success that underground dinners in the Maori became part of the Gympie social scene, and some riotous revels resulted.

The owners used to recall the time they entertained Paddy Perkins, who had just started a brewery in Brisbane. Paddy enjoyed his dinner so well that he became boisterous, and the only way they could get him back to the surface was to tie him into the bucket. By the time everybody else was brought up the party had become hilarious, and the next thing Paddy knew was that his bonds had been cut and he was decanted from the bucket to go rolling down the side of the mullock heap.

With every local stream polluted by the tailings of the stampers, an adequate supply of fresh water remained one of the town's main problems for years, though many of the miners were hampered by striking more than enough of it in their shafts. A story was told that one night when diphtheria was raging in the town, one of Ted Poulton's sick children woke up calling for water. The makeshift tank was dry and Poulton grabbed a rope and a bucket and went out to an abandoned shaft which he knew was flooded. As the bucket touched bottom a distinct groan came up from below. It was midnight, but Poulton dropped the bucket and ran for help, returning with constables Hazlett, Currey and King, all sleepy and none too happy at having been disturbed.

Timber, ropes and tackle were brought and one of the policemen was lowered thirty or forty feet into the shaft. He returned, grim faced, to announce that it was not a man down the shaft, but a cow. By this time half the town's population was out of bed and gathered round the shaft, and with a lot of willing help the cow was eventually hauled to safety. Con Toomey's cow had been missing for a couple of days, and having heard what was happening, Con arrived on the scene to claim her armed with a bottle of whisky for the rescuers. The whisky, at least, was welcome, and it had all disappeared by the time Con discovered it was not his cow at all, but another animal belonging to Jerry Keliher.

In spite of phenomenally rich yields in the first five years, there were indications, as the early 1870s advanced, that Gympie faced a period of decline. The shallow reefs were being worked out and only a few diggers had the capital and the doggedness to go deeper. Many were recalling geologist Aplin's report of July 1868 in which he had described the rocks of the Gympie area as consisting of slates, sandstones, grits and conglomerates associated with a massive greenstone, and the whole traversed by quartz veins of varying thickness of a generally gold-bearing character.

Aplin had pointed out that the greenstone occurred at intervals over a large part of the goldfield, and that it was in the decomposed upper portions of this rock that the quartz veins were found to be so productive. In its ordinary condition the greenstone was excessively hard and a most formidable obstacle for miners to drill through. He had gone on to say:

> From the crystalline character and extreme hardness of the greenstone there is little possibility that the reefs traversing it can be profitably mined to any great depth. Though they may not be altogether pinched out, the quartz veins will doubtless become so attenuated that it will no longer be profitable to work them.
>
> Such is the case with metallic lodes generally when associated with greenstone or other crystalline rocks of like origin and composition. They are frequently very rich as long as the rock composing them is in a decomposed condition, but invariably become poorer and reduced in size when they pass below the depth to which decomposition has reached.

As early as January 1869 many thought they saw indications that Aplin was right and that the gold would not continue to any great depth in the crystallized greenstone.

Together with declining hopes for the Gympie reefs came news of a succession of new goldfields being opened in the north of the colony—Gilbert River in 1869, Ravenswood in 1870, Etheridge and Charters Towers in 1872. Men who had been beaten in the battle against the tough greenstone rolled their swags and headed north.

Then, in 1873, a man Gympie miners knew well, their old mate James Venture Mulligan, found gold on the Palmer River. "It's a river of gold," he was reported as saying. The trek out of Gympie became an exodus. Not only diggers, but storekeepers, publicans, shanty-keepers, and a whole horde of gold-town hangers-on sold out or walked out and took ship north to the Palmer. With them went French Charley Bouel, to trade good wines and gay girls for Palmer gold on Cooktown's muddy shore.

Through all the vicissitudes of Gympie's hard years, one man's phenomenal luck did not desert him. Lucky Leishman, the former sailor who happened to be in Maryborough at the right time and meet the right man to take him as a member of the first party on the field, never made a bad mistake.

After the best of the alluvial gold was finished and Nash had sold out and left on a trip to England with his wife, and Billy Malcolm had sold out and returned to his native Scotland, Leishman remained on the field and used some of his gold to buy the Wheatsheaf Hotel half-way up Caledonian Hill. With a steady income assured, he tried his luck in a number of reefing claims and eventually, in 1873, became interested in the old No. 7 and 8 South Lady Mary claim on which a shaft had been put down about sixty feet and abandoned.

Leishman could not work the claim alone and looked around for a partner. For a long time he could not find one. The first flush of reefing enthusiasm had subsided, outside capital was scarce, and mining was mainly in the hands of groups of experienced local men who knew the ground well and invested their money and their labour shrewdly. A sixty-foot shaft on an abandoned claim was not good enough for them.

Leishman persisted and at last persuaded three men—Julian Brier (known as Jack the Frenchman), Ignatius McMahon, and G. Duckworth—to take a half share between them while he took the other half. The four of them soon got the old shaft cleaned out and began digging. They worked in two shifts, Brier and McMahon on one, Leishman and Duckworth on the other.

Mrs Leishman's fourth child was about due to arrive, and when the time came, Leishman, leaving Brier and McMahon on the night shift, brought Doctor Ryan and then spent the greater part of the night pacing the floor of their tiny, slab-walled living room.

Bringing in the Big Cake | 165

Sometime in the small hours of the morning the doctor appeared in the doorway of the bedroom smiling broadly, to announce that Leishman had another son. Out of the flurry that followed the doctor was given a drink, and was making for the door when it burst open in his face to admit Jack the Frenchman, eyes shining, arms waving, and speechless with excitement.

"Good heavens man, what's the matter?"

"Gold," yelled the Frenchman. "Gold. The mine it is full of gold. There is gold everywhere. We are all rich, rich, rich."

He embraced Leishman, he embraced the doctor, and he was headed for the bedroom to tell Mrs Leishman when they stopped him and explained.

As things quitened down they went and told Duckworth, and then they all ran back to the mine, where McMahon had been keeping watch.

It appeared that Brier and McMahon had lit the fuses of the charges in two drill-holes before coming up for crib. They heard the charges explode, and when they went back down the mine there was gold all over the bottom of the shaft.

The four partners remained down the mine all the rest of the night collecting specimens. There was a lot of jagged chunks of pure gold and everything they picked up seemed to be more gold than quartz.

In the morning they chipped the pieces of quartz off the gold and carried it all straight to Mr Joseph, the assayer. They watched it smelted, and while the bar of solid gold was still too hot to pick up they doused it with a bottle of champagne to cool it. Then they picked it up between them and, followed by a mob of cheering miners, marched with it in triumph to the Bank of New South Wales.

The first crushing of ore from the mine was put through the big New Zealand battery which had lately been erected by the owners of the Maori mine. Young Billy Leishman, who was watching, recalled later that the ore was so rich that masses of solid gold jammed between the stampers and had to be cut away with a chisel.

Good crushings continued. McMahon sold his share to Brier and Duckworth. Leishman promised his wife that when they had made five thousand pounds he would take the family home for a holiday to Scotland. The five thousand pounds was soon made, a company was formed with Leishman still holding a substantial share in the mine, and they left for the promised holiday.

While in Scotland they renewed the friendship with the man who had put them on the trail to wealth—their old mate Billy Malcolm, who had used his Gympie gold to buy a hotel in the Lakes District.

By the time the Leishman family—increased by yet another son—got back from Scotland the mine was nearly worked out. Lucky Leishman, who once again had arrived at exactly the right time, sold his shares in it and, still not content to sit back and enjoy the fruits of his labours, headed north.

He had another look at the Burrum coalmines, now prospering as suppliers of coal to the Gympie batteries, but after the goldmines coal had no attraction for Leishman, and he continued his trek north, prospecting as he went. As anybody who knew him might have expected, he soon struck gold again, this time among the hills to the south of the old Crocodile Creek diggings.

There was a popular play on tour at the time called *Struck Oil*, and as the new gold find crushed at twelve to sixteen ounces to the ton Leishman called his new claim Struck Oil. A township of that name soon grew up around it.

Leishman died at Rockhampton aged eighty-six, a very wealthy man and lucky to the last.

Chapter Eighteen: Black Rebel

By the time of the Gympie gold-rush the Aborigines of the area had already lost their country to the squatters and timber-getters. They had acquired a taste for sugar and rum and were prepared to do a certain amount of work or hire out their women to get money to buy these luxuries. Outlying tribes flocked in to the fringes of the diggings to be closer to the source of supply.

Diggers employed the men to clear timber and build huts, shanty-keepers traded them spirits. When they became drunk they generally quarrelled among themselves, and then they were driven out of town.

In their original state they were a well-built, sturdy race with a natural grace and proud bearing. Their country was rich in such native foods as berries and nuts, there was an abundance of native game, and the river, as one digger put it, was "so full of mullet that even white men could catch them". Men and women alike went naked, and in the early days of the diggings nobody minded. The few white women on the field were familiar enough with life in the bush to take it as a matter of course.

As families moved in the Aborigines who came into the township were given cast-off clothing and were chased out of town if they appeared without it. Dress for a man was sometimes a pair of old moleskins, or just as often an old shirt with a belt or piece of string round the waist. Women's dresses were voluminous neck-to-ankle garments which were sometimes given a more interesting line by Aboriginal women by one arm being thrust through a sleeve, the other through the placket opening at the waist.

Before long all these people had been concentrated in a few camps outside the town.

For centuries the area had been the traditional meeting place of coastal and inland tribes for their formal battles to settle differences, for corroborees, councils and initiations. These meetings continued through most of the gold-boom period and were a source of entertainment to the townspeople and surrounding station workers who, on special occasions, would ride in from a hundred miles or more to watch.

Early one morning in 1872 the whole town stopped work to watch a big fight between nearly two thousand inland Aborigines and about an equal force of coastal, Kabi tribesmen. The warriors met on one side of a ridge and battled it out with spears, shields and nulla-nullas while, over the crest, the women fought their own battle with digging sticks.

All morning and well into the afternoon both battles raged before growing crowds of barracking whites as word of the conflict spread and men came riding in from outlying areas. It was getting on towards sunset, with one warrior killed in the men's battle, and a woman killed in the other, before a truce was called and the former enemies at once mingled like the best of friends and began a corroboree that lasted for more than a week.

As farm-houses spread over the fertile flats of the lower Mary River, it became common for Aborigines to come in asking for food and tobacco, it being understood that before doing so they should lay down their weapons some distance away as a token of good faith. In many cases friendly relations were maintained between tribesmen and settlers, white and black children played together, and some semblance of mutual respect developed. But no matter how friendly the two races might become, brutality and intolerance were never far beneath the surface.

The old Kabi pride of race and bitterness over lost hunting grounds occasionally found expression in a refusal to accept white men's rules, and the most easygoing white settler was always on the alert to see that a "cheeky blackfellow" was put in his place. A white man had only to reach for his gun to send any Aboriginal bolting for cover.

Mrs John Gillis, of Goomboorian, about fifteen miles north east of Gympie, recalled that one day when the men were away an Aboriginal came up from their camp and strode into the house carrying all his weapons. Mrs Gillis turned on him and demanded: "What for you fellow bring them inside?"

Ignoring the question, he asked for tea, baccy and flour. Mrs Gillis crossed to the open window shutter to see if there was any sign of her husband returning. As she did so, in her nervousness, she picked up the open case of her husband's big meerschaum pipe and snapped it shut. The man, apparently thinking it was a revolver being cocked to shoot him, bounded out through the door and disappeared.

When Mrs Gillis reported the incident to her husband he grabbed up his bullock whip, rode out to the native camp, and threatened to flog everyone there unless they told him who the culprit was. They pointed out the man, and Gillis flogged him until he could not stand.

There were cases of lonely diggers being surrounded by Aborigines and robbed, but there were few instances of spearing or deliberate ill-treatment.

One digger was released after being stripped of everything he had, and a couple of days later staggered up to a station homestead, exhausted, naked and plastered with mud, which he had put on to protect his body from the sun, sandflies and mosquitoes.

A cedar-getter named Frank Luck engaged a Gympie Aboriginal to show him a stand of cedar further up the river. On the way they stopped at the Seven Mile Hotel, on the Brisbane road. Luck bought the Aboriginal a drink and, with the man watching him, changed a sum of money. When they were well into the bush Luck heard a low whistle and looked round to find himself surrounded by Aborigines, one of whom was in the act of launching a spear. Luck shot the man in the hand with his revolver and then pointed the weapon at his guide and threatened to blow his brains out if the man did not get him out of the place safely and quickly. The surrounding tribesmen vanished and Luck escaped.

Farming developed fairly rapidly around Gympie during the 1870s, and as all the good agricultural land was taken up, Aborigines were crowded more and more into a few big camps around the town. Soon they became, in the words of the townspeople, "a nuisance".

Mobs of hungry Aborigines would assemble on the fringes of the town, spread out, and work their way right through it, begging from door to door. Mounted police regularly herded them out, but in a few days they would be back doing the same thing. They began sneaking into town at night and raiding beehives and toolsheds. A number of "smoking out" expeditions were organized against their camps.

Walter Ambrose, at one time Mayor of Gympie, recalled that as a boy in the 1870s he and some friends held Constable Downey's and Constable Martin's horses while the police set a whole camp of bark gunyahs alight and burnt it out completely. For days afterwards, while the Aboriginal men were at work building a new camp, the women were fossicking about the old one for anything that had survived the flames. They were stark naked and they carried babies and puppies wedged under their arms as they worked.

The end of the tribes was inevitable, but the story of their decline is highlighted by the exploits of one remarkable man —a sort of Aboriginal Ned Kelly, said to have been driven to crime by unjust treatment—who terrorized the district for nine months before being hunted down. He was a young Kabi tribesman named Kagariu, better known by his white man's name of Johnny Campbell.

Many of the stories recorded of Campbell are vague, and not all versions agree. A few white settlers of the time said he was started on his career of crime by being blamed for an assault that he did not commit. Others said he was led on by a white girl who later complained of his attentions. A story was told that he waylaid a schoolmistress riding through a property on which he was employed, that she slashed him across the face with her whip and later complained to the manager who had Campbell tied to a tree and flogged.

Young Johnny Campbell was certainly a lad of more than average intelligence. He learnt to speak fluent English and received some education in a Sunday school conducted on the station of one of his early employers. He grew into a first-class bushman, learnt to ride "better than the average white man", and became a deadly shot with a rifle.

He was described by E. Forman, of Esk, who employed him as a shepherd and horse-breaker as "an average type of blackfellow, and a good and conscientious worker". Forman frequently had to leave his seventeen-year-old wife alone on the holding with Johnny, and there was not any hint of his attempting to molest her in any way.

After Johnny left Forman something went wrong. Soon after that he went bush, riding a stolen horse and armed with a stolen carbine. There followed a number of assaults and robberies for which Johnny Campbell was blamed, and in 1872 he was charged with assault, found guilty, and sentenced to seven years' jail.

Whatever he may have been before that, Campbell came out of jail in 1879 embittered, savagely resentful, and determined to revenge himself on the whole white race for whatever had happened to him. He told a white settler soon after his release that after what he had seen white men do to Aboriginal women he knew that the black man could never get justice, so why should he try. From then on, apparently, he never did try.

He took to the bush in the Kilkivan district, and soon a whole series of thefts of food, clothing, arms, ammunition and other articles was attributed to him. It was reported that he and another Aboriginal, both armed, had held up and robbed the occupants of two isolated farm-houses. Then, about eleven o'clock on the night of 19 June 1879, a ten-year-old girl screamed that there was something in her room. Her awakened parents rushed out in time to see a dark figure vanishing into the night.

Next day three houses at Mount Coora, about ten miles to the south of Kilkivan, were robbed. Other robberies followed, and by the middle of July a large force of police was out looking for Johnny Campbell.

Campbell knew all the Gympie country thoroughly. He was a man of remarkable agility and physical endurance, and in bushcraft knew all the tricks of both black men and white. Though he could easily have stolen all the horses he wanted, he chose to travel on foot, mainly at night, slipping through his familiar bushland without leaving a trace. His pursuers rarely so much as sighted him.

Then, soon after midday on 3 August, an Aboriginal woman named Mary Ann came into the police barracks at Kilkivan and complained that the previous night Campbell had come into her hut on a near-by sheep station, grabbed her by the arm and told her to come with him. When she resisted he knocked her down, tore off her clothing, and was continuing the attack when her screams brought other Aborigines to her help. Campbell then ran off.

Meanwhile, a station Aboriginal named Big Toby had told the station owner, W. Ross, that Campbell was in a near-by Aboriginal camp and had given Toby money to buy him a bottle of rum at the Rise and Shine Hotel. Ross mixed a potent brew of wine, beer and brandy, and handed it to Big Toby with the promise that there was five pounds reward for him if he could get Campbell drunk and grab him. Toby headed into the bush, and Ross and a couple of station hands, armed with rifles and on foot so as to make as little noise as possible, followed close behind.

As they came in sight of the camp somebody warned Campbell, who came charging from a bark shelter straight towards them. Sighting them only as one of the station hands raised his rifle to fire, Campbell bounded aside in time to dodge the bullet and bolted to the cover of a near-by creek. By the time the pursuers came up he was gone. They could find no sign of his tracks and, though threatened with flogging, none of the tribesmen would admit that they could see any either.

Six days later Campbell rode up to a native hut at Cinnabar, about four miles south-west of Kilkivan, on a stolen chestnut station horse, seized a twelve-year-old Aboriginal girl, swung her onto the saddle in front of him, and rode away, felling with a blow of his fist an old woman who tried to stop him. The horse was later found in the bush unharmed, but of Campbell or the girl there was no trace.

About a week later, in broad daylight, Campbell came to the door of a house about ten miles out of Gympie, told the woman who was its only occupant that he would blow her head off with his shotgun if she tried to stop him, helped himself to food and her husband's watch, and left.

A few weeks after that a woman alone in a farmhouse near Kilcoy, about fifty miles to the south, saw a black man stealthily approaching the building. She slammed and barred the shutters and door and grabbed a loaded gun kept always handy for emergencies. Campbell, abandoning his attempt at concealment, strode up to the door, hammered on it with the butt of his gun, and threatened to shoot the woman if she did not let him in. She threatened to shoot him if he did not go away. The stalemate was ended by the sound of horse's hooves as the husband returned. Campbell vanished into the bush.

The following day Campbell watched another farm-house a few miles away until he saw the man leave, and then strode boldly up to the front door and threatened to blow the woman's brains out if she did not stand aside. She snatched up her infant child and ran. He made no attempt to stop her, but ransacked the house before leaving. Local police and trackers were scouring the whole district but they found no trace of Campbell. He was apparently on foot and travelling mainly at night. It was learned only later that one of his tricks in settled areas was to walk along the top rails of the post-and-rail fences so as to leave no tracks.

At this time Campbell had two native women with him, and on a number of occasions he sent them in to farm-houses to ask for food and tobacco. During almost the whole time he was terrorizing the district he had at least one Aboriginal consort, and before entering an Aboriginal camp he would always send her ahead of him to find out if he would be welcome or not. He never showed himself until he was sure. He was frequently accompanied on his travels by a sort of Aboriginal bodyguard which slipped through the thick bush ahead of him so as to be able to give the alarm if anybody was waiting in ambush.

As the hunt continued, however, Campbell gradually lost the sympathy of his own people, partly because of constant harassment by police search parties, partly because of his own habit of discarding his consorts as soon as he was tired of them and stealing from her tribe any woman who took his fancy. One of these stolen women showed herself not quick enough to please him and he brained her with a nulla-nulla.

By October 1879 Campbell was back in the Gympie district, hiding out in bushranger Palmer's old haunts—the thick scrubs of Eel and Pie creeks—robbing lonely houses while the menfolk were away and vanishing before pursuit could be organized.

He was nearly caught while robbing a timber-getters' camp at Amamoor Creek, about ten miles upstream from Gympie, and on 31 October tried his hand at bushranging by holding up a bullock-driver and a digger at Ringtail Scrub between Gympie and the Noosa River.

Constable Tom King, of Gympie, one of the best bushmen in the area at the time, and a black tracker followed the trail and came on Campbell unawares in a bush pigsty where he was hiding. King pounced on him, but after a long struggle Campbell came out on top. He snatched up his tomahawk and would have split King's skull with it had not the tracker grabber his arm. Even then Campbell managed to break free from both of them and make his escape.

But the pressure of the pursuit was beginning to tell, and the local Aborigines became daily less inclined to help. Campbell headed inland with his latest consort, a huge Kabi woman said to have weighed about eighteen stone. A reward of fifty pounds was out for his capture. On 5 January 1880, at Neurum Creek, a few miles east of Kilcoy, running true to what had become his pattern, he watched a farm-house until he saw the man leave, threatened the woman with his gun, and took what he needed.

A few weeks later, on 10 February, he crept up on a farm-house at Kipper Creek, about thirty miles south of Kilcoy, and watched until he was sure all the men had gone. The farmer's wife later gave evidence that her husband had gone out looking for the horses, leaving her and her fifteen-year-old sister alone.

About 9 a.m. Campbell came to the house and asked for matches and then leaned against a wagon talking to them about the blacks. She said it was raining and she gave him a sack to put over his shoulders.

Suddenly he produced a revolver and made an improper demand on the young girl. The elder sister called for help, and he pointed the revolver at her head and told her to shut up. The young girl ran from the house; Campbell followed her, overtook her about twenty yards away, threw her to the ground and assaulted her. He then disappeared into the scrub. When the two sisters were sure he had gone they ran to a house about two miles away for help.

Campbell himself denied this version of the story. He said he had never threatened either of the women, and that the younger one had deliberately enticed him into the scrub near the bank of the stream.

In the hue and cry that followed the incident, all available police and trackers were called out, among them a Sergeant Campbell and his black tracker, Billy. Billy followed Campbell's trail back to Neurum Creek, and he had just told the sergeant they were getting close to their quarry when he thought he heard a twig crack.

Without a moment's hesitation he leapt from his horse and, carbine in hand, plunged through the scrub towards a big fallen tree from which the sound seemed to have come. The next thing he knew he was looking down the barrel of a rifle. There was a flash and a roar, and Billy reeled back with a bullet through his left shoulder.

A black figure rose from behind the fallen tree and vanished into the scrub. The sergeant was off his horse in an instant and fired in the direction of the vanishing fugitive. He followed a short distance and then stopped to listen, but there was not a sound. To go further would mean certain ambush. He returned to the injured tracker.

Billy was badly hurt. The two of them headed for the nearest homestead. On the way they met Constable Tom King and his new tracker, Johnny Griffin.

Griffin himself was a Kabi tribesman who had fallen foul of white man's law and been charged with the murder of a squatter named Stevens. He was acquitted, and the police, deciding they might as well make use of his abilities, enlisted him as a tracker. This was his first patrol and neither Campbell nor any of his own tribesmen knew he was now working for the police. He was, by a coincidence, the half-brother of Johnny Campbell's eighteen-stone consort.

Griffin made up his mind from the start that he was the man who was going to catch Johnny Campbell. From then on he stuck to the trail like a bloodhound. Day after day he followed Campbell's tracks over country that both of them knew thoroughly. Often he lost the trail—sometimes for a day or more—but, unlike other trackers, he never gave up. He simply circled round patiently until he found it again.

In an effort to shake him off, Campbell and his consort doubled back and forth, waded up creeks, followed stony ridges and laid false trails, but in the end his unknown pursuer always dogged his footsteps.

Ten days after the shooting of Billy, Campbell, hoping to find country he knew better than his pursuer, headed back for his home territory at the mouth of the Noosa River.

It was a hard trek that led far up into the watershed of the Mary River and over untouched mountain country east to the coast. Here, he felt confident, he could not be followed. What he did not know was that his unseen pursuer was a Noosa man as well.

As it became clear to Constable King where their quarry was heading, he left Griffin to follow the trail and made his own way by the roundabout but easier route to the old logging camp of Tewantin, a few miles up-stream from the Noosa River's mouth. From then on Campbell might have been followed by a ghost; not a sign did Griffin give of his presence. The fugitive became confident that he had at last shaken off the pursuit.

Slowly Griffin closed in until, within sight of the camp at the mouth of the Noosa, he circled round so that he and Campbell approached it from different points of the compass. They arrived within a day of each other, both naked and carrying nothing to identify one as a fugitive, the other as hunter.

Campbell followed his usual practice of sending his consort in ahead to test his welcome. She reported it was safe. His reluctant hosts knew Campbell had a price on his head and they wanted no trouble with the police for sheltering him, but on the other hand Campbell was a dangerous man to cross and no one was inclined to oppose him.

When Griffin arrived he told the tribesmen he had been working on a station and was on walkabout. He pretended he did not know Campbell, and for a while was content to sit back and watch him, biding his time. Campbell, though suspecting nothing, remained watchful, carrying a loaded carbine wherever he went, and his consort was never far away with his revolver.

Accounts of how Campbell was captured differ in detail. Griffin at last persuaded some of the bolder warriors that they should seize him while he was asleep and hand him over for the reward.

They waited their chance and rushed Campbell while he was off guard. When some timber-getters arrived in response to a signal on 15 March 1880 they found him bound up like a cocoon with a clothes-line, still struggling and yelling threats at his captors. When King saw him he put on a pair of handcuffs as well.

On the way to Maryborough for trial Campbell was lodged overnight at the Gympie lock-up. The whole town turned out to see him arrive. When he was taken off the Cobb and Co. coach he was still handcuffed and bound hand and foot and all round his body with the clothes-line. Even so, an armed policeman had travelled on either side of him in the coach while four mounted troopers rode escort, one on either side, and two behind.

On the box seat outside with the driver rode the huge Aboriginal woman—"looking about six feet wide" according to one digger. She remained as close to Campbell as she could all through his trial at Maryborough.

Early in April 1880 Campbell was sentenced to fourteen years' imprisonment on a charge of assault and robbery. After serving three months of the term in Brisbane jail he was brought before the Criminal Court at Ipswich and charged with the rape of the fifteen-year-old girl at Kipper Creek. He was convicted on 26 June, sentenced to death, and hanged at Brisbane on 16 August 1880, unafraid and defiant to the last.

The Noosa men who had captured him were rewarded with the gift of a whaleboat and fishing nets.

Chapter Nineteen:
"Jeweller's Shop" Days

The five years the northern rushes stole from Gympie's gold boom saw many changes. Miners who came trooping back towards the end of the 1870s found not the ghost town some had expected, but a tight little community which had closed its ranks and battled through to the threshold of a new promise.

The town itself was much smaller, the tents and most of the bark huts were gone, and the buildings that were left had a substantial and permanent look. Most of them were of timber, a few of brick. On the flat, where ten years before the Chinese had been washing out the last of the alluvial, a large sawmill now ripped through a brisk business.

A few mines—notable among them No. 7 and 8 South Monkland, No. 2 North Lady Mary, Glanmire P.C.—had continued to produce good returns, while in others men who refused to be discouraged by gloomy predictions, maintaining that a man swinging a pick and shovel was the best geologist, slogged on to force a passage through the tough greenstone to the gold they were convinced lay beneath it. Mining methods had improved a good deal, but mechanization had hardly begun and most of the stone was still raised by horse-drawn whims.

Miners who were getting gold, and members of the business community who had remained on, realizing that the future of the field depended on new gold finds, had bought these companies' shares and paid calls for capital with dreary regularity, even though the chances of ever getting dividends from them often seemed slender.

Down in the mines the men worked on steadily until, beneath the greenstone, they came once again on slate. It was darker and harder than that through which the surface quartz veins had run, but it was similar. Wild excitement swept the field, and the red flags were flown as though gold had already been found in the promising new beds.

Progress was slower in the deep mines than it had been near the surface, but once on the slate, gold was soon found and the year 1880 opened on a wave of optimism. Month by month news from the mines laid the foundations of the new prosperity.

> GYMPIE, January 23: In a great measure owing to the more systematic method of mining being pursued of late years and the revolution which has been wrought by the use of new explosives and small steel, but more on account of the more extensive knowledge gained with regard to our quartz lodes, several reefs—which, in the years after the discovery of the field, were prospected, a large amount of undirected work expended upon them, and then duffered out—are again being occupied and are bringing in remunerative, and in some instances, rich returns.
>
> April 2: In the Phoenix Company mine things are beginning to look up and auriferous stone has been met with at the 404 foot level.
>
> April 23: Good yields are being looked for from some mines the owners of which have been prospecting at lower levels. One of these is the North Phoenix Company.
>
> May 7: At a general meeting of shareholders of the North Phoenix Company's mine, it was decided at once to proceed with the erection of steam machinery and winding gear.
>
> May 14: The third and largest escort of the present year left on Wednesday morning taking 6,127 ounces. The larger half of the parcel is the produce of the North Glanmire and Phoenix mines.
>
> June 11: All the batteries are in full swing. The North Glanmire, No. 7 and 8 Monkland and No. 1 North

Glanmire each produced a large quantity of stone and before they are finished, new stone from other mines will be forthcoming.

July 16: The principal event of the week has been the coming upon splendid gold in the Glanmire Prospectors' mine in the breaking down of stone in a winze [a sloping passage connecting two levels] sunk from the 300 foot level. Specimens estimated to contain fully 200 ounces of gold were obtained.

Then, on 5 December 1880, came the discovery that began Gympie's new boom. As had happened time and again on the field, it was a result of faith and perseverance.

The latter months of 1880 had seen the main vein of quartz in the No. 1 North Phoenix gradually petering out. The shareholders, already committed to the expense of installing steam winding machinery, found the spectre of bankruptcy looming closer almost day by day. At the 332-foot level crosscuts were driven through the rock in search of fresh quartz without result. Disagreements arose among the shareholders and some sold out. Others refused to give up. They maintained that the true Phoenix reef had never been reached in No. 1 North Phoenix, and that what they had been working was either a continuation of what was known in the original Phoenix mine as the Western reef, or else a branch of the main reef.

More than three hundred feet underground, first in one direction, then in another, the search for the main reef continued, every foot of it hammered through hard rock. Towards the end of the year, with the remaining shareholders' money nearly gone, work was resumed on the eastern crosscut at the 332-foot level. Thirty feet of rock was dug out with no promise of gold—the next shot fired showed a two-foot-thick vein of quartz gleaming with gold. Within twelve months the once almost bankrupt shareholders had divided £64,000 between them.

By 1882 the steam winding machinery had replaced the horse-driven whim at the No. 1 North Phoenix pit-head, and before long the company's stamping battery had grown from fifteen heads to sixty.

The North Phoenix success brought capital flowing back for the deep mines. Abandoned claims were reopened and their shafts driven down to levels once believed to be barren. Gold was found in greater profusion than ever before on the field.

When considering values of early gold production, one has to keep in mind the fact that during the latter part of the 1800s the purchasing power of money was generally from six to eight times what it is today. On the other hand, the price of gold was much less, ranging from about £3/5/- to £4 ($6.50 to $8) an ounce compared with about $36 an ounce today.

In mid 1883 Wilmot Extended mine, on the north bank of Deep Creek, struck paying stone, and in September paid a dividend totalling £4,500 from its first crushing of 1,400 ounces. The reef improved as digging went on, and some of the first ore brought up in 1884 showed as much gold as stone. A crushing of 231 tons of ore returned 4,549 ounces of gold worth £16,173.

Day by day excitement on the field mounted as one rich patch after another was uncovered. Then in March came the crushing that set Gympie's record and began what came to be called the "jeweller's shop" period. In thirty hours 470 tons of ore went through the stampers to yield 10,944 ounces—more than a third of a ton—of smelted gold.

The resultant dividend to shareholders was one pound a share, or a total of £36,000. Individual dividend cheques for amounts up to £4,702 and £4,525 went out to men who, twelve months earlier, had hardly a penny to their names. For the twelve months ending September 1884 Wilmot Extended dividends totalled £99,900.

Alongside the Wilmot Extended mine was the Ellen Harkins, which bottomed on rich stone at 572 feet in February 1884, and from then on every foot the mine went down uncovered richer ore. There were places where the miners found themselves literally driving their drills into solid gold.

In six weeks the four men sinking the shaft took out £15,500 worth of gold within a space of twenty feet. A crushing of seventy tons nine hundredweight yielded 4,468 ounces of gold, and dividends totalling £14,850 were paid to shareholders who had recently been describing themselves as "the poorest devils on the whole goldfield". The next crushing was fifty-three tons sixteen hundredweight of ore for 2,506 ounces of gold. The year's output for 2,177 tons of ore was 14,728 ounces of gold.

On the Monkland ground the pioneer deep mine was the Great Eastern. Many had regarded it as a "wildcat" mine from the start, and as years slipped by without results they seemed to be justified. Then in 1886 the mine got onto gold. The country was soon pegged for miles around by those who had doubted, and ground was broken on claims that were to become famous as No. 2 South Great Eastern, South Glanmire, and the Scottish Gympie.

With mine after mine breaking into the big-dividend bracket, the Gympie Stock Exchange was opened on 10 July 1884 with 127 members and sixty companies listed. An orgy of speculation followed, and fortunes were made and lost by men who had never been near a goldmine in their lives.

As distinguished visitors began to come to the field to see what a goldmine looked like, No. 1 North Phoenix—under the management of Mr George Argo and employing about 120 men—became one of the show-places. A description of the mine given by Aleck Nimey in his *Gympie Mining Handbook* of 1887 is typical of the more advanced Gympie mines of the period.

> In Gympie's centre, after you have surmounted the Caledonian Hill, you find on a gentle descent on the other side the big, red brick chimney of the No. 1 North Phoenix with an enormous galvanized iron shed alongside in which the music of sixty stampers is, like Tennyson's brook, for ever and ever. There is an engine house too, alongside, out of which the two iron ropes come that pull the cages up and down the main shaft.
>
> A pagoda-like structure known as a poppet head stands just over the divided shaft, in which revolve a couple of immense wheels over which the steel ropes go continually, the directions of the engineer being given from below by signal. The cages at the end of these ropes will take trucks and travellers with equal ease.

There is just room for four people in the descending cage in which we are soon gliding down in the semi-darkness, with miners' clothes on our backs and miners' candles in our hands. When you add the ancient slouch hat and heavy boots sufficiently large to turn round in, you are ready for anything. As you step in the cage you hear the signal given to the engineer as to the particular level you want to be deposited at.

There are four such levels here, at 332 feet, 428 feet, 500 feet and 580 feet, the shaft going yet deeper in order that the waste water may be the easier pumped out.

At the 332-foot level there is a sort of platform around the shaft and by the light of a candle or two we can see two dark abysses leading both ways, north and south. In the centre of the dark avenue, or level, as the miners call it, is an iron tramway sixteen inches wide which is overhung at regular intervals by shoots in which are stored large masses of quartz put in at the overhead levels, and which can be discharged into the trucks on the tramway by just lifting a lever.

Let us go down one of the winzes, or shafts connecting the different levels, where the reef has been traced down, down, down, to the very bottom of the mine.

A few descents on iron-runged ladders bring us to the 500-foot level where both the Phoenix reefs have spread out in a mighty mass overhead and underfoot to varying widths of twenty-nine, thirty and even forty feet. There are men here and there at work with candles stuck on some rocky protuberance handy, who keep on at their labour of drilling and blasting sixteen hours, all night long, in the usual two shifts of eight hours each.

As well as we can see in the semi-darkness, the roof is very variable, sometimes breaking out in white streaks or leaders, and sometimes intruded on by great masses of mullock or worthless stone that has to be sorted afterwards from the reef stone. As a rule, however, it pays to crush nearly everything as the sixty iron stampers that make the eternal "music of the mine" overhead can crush as low as five pennyweight to the ton and then make a profit.

Further on the two reefs divide again and the open space gives way to two tunnels that in time unite again the lower down we go.

Above this level are four winzes to the north of the main shaft and three to the south, all on the reef, following it on the skew or underlay. Above the lowest, or 580-foot, level there are five winzes to the north and three to the south, making a perfect honeycomb in this great Tom Tiddler's ground.

As we proceed in our subterranean wanderings we are shown, here and there, little spots of yellow metal in the reef, some of it a dull yellow and some shining out where it has been severed from other gold.

As far as the Phoenix No. 1 is concerned these little discoveries are not allowed to be worked on exclusively, as the great majority of the stone is under the average ounce to the ton, and the best is carefully admixed with the middling, with the result that the returns show very little, if any, fluctuation.

All the Gympie claims have not been so prudent, and some claims which made a tremendous spurt in 1884 are now suffering from dead work. At any time, should it be necessary for the Phoenix No. 1 to spend large sums of money in non-paying toil, the good stone will still be there to fall back upon to keep up the average returns and good name of the mine and the value of its shares in the market.

It is a thing that Queensland miners have to be thankful for that their lives are taken much more care of than those of the average sailors at sea. The laws are very stringent, the consequence being that out of a field of about 11,000 persons only three miners met with fatal accidents in 1885 and thirteen received injuries.

Every big mine is required to have the passages timbered as far as they are excavated. In the No. 1 North Phoenix the old timber, consisting of ordinary blocks of timber filled in with ordinary boarding, is being replaced with enormous beams which rise on either side like the rafters of a house and meet in the centre of the top of the drive so that the greater pressure there is from above, the greater is the resistance below.

The No. 1 North Phoenix battery at this time consisted of sixty stampers driven by three separate steam engines and working continuously to put through one hundred tons of ore each twenty-four-hour day—or six hundred tons in a working week. Gold recovery appliances were capable of obtaining 97 per cent of the gold in the ore.

Before the gold gave out, No. 1 North Phoenix, the mine whose owners had driven on down into the deep rock after others had given up, had produced 248,877 tons of stone which was crushed for a total of 238,048 ounces 6 pennyweight 2 grains of gold, worth £848,678/17/11, from which dividends totalling £436,750 were paid.

Chapter Twenty: The Great Flood

The first white settlers of the Mary valley had been cedar-getters, and they continued their work throughout the gold-rush until, towards the end of the century, there was hardly a good-sized cedar tree to be found. Thousands of feet of timber were sent south to the builders and furniture-makers, much was squandered locally on farm buildings, and even more was left to rot in the forest.

Vast stands of huge cedar trees were felled, their trunks branded with their cutters' marks, and snigged by bullock teams to creeks and gullies and left there for the summer floods to carry to the Mary, and so down-stream to the broad reaches of the river where they could be bound together into rafts to continue their journey.

Often the creeks were badly chosen, and their wet season waters were not enough to carry away the logs which then lay there year after year until they rotted away, while the cutters, rather than snig them out, hacked their way on into the virgin forest.

Trees felled closer to the river were dragged to its steep banks and rolled down into the water to be carried down by the current.

The Mary River was normally in flood from January to March, and by the beginning of every wet season its upper reaches were jammed with logs waiting to be "freshed" down. The cutters followed them down in boats to free the jams that often occurred. The operation was hazardous enough, but less so than one of the popular amusements of the youth of the goldfield who, when the logs were coming down, were in the habit of hiking up-river and then swimming out, each to mount a likely looking log and see how far he could ride it downstream. A few crushed legs caused by colliding logs did little to deter them. Dozens of snakes came down on the logs, and these were disposed of by the swimmer rotating the log in the water a few times. The dispossessed snakes could swim every bit as well as the boys, and the possibility of their attacking their tormentors in the water was accepted as part of the game.

In Gympie itself the river was often jammed so solid with thousands of logs of cedar, up to five feet in diameter and a hundred feet long, that one could easily cross the stream on them, and they also became a popular playground for the local boys.

At Tiaro the tidal water began and farmers made extra money in the wet by going out in boats, snaring the logs and bringing them in to the banks for rafting, which consisted of binding them together with chains passed through steel dogs, or looped spikes. The cedar-getters paid a shilling for every log recovered at Tiaro, and more for those caught further down-stream, but in spite of this hundreds of logs were swept out to sea every season and never recovered.

Although there were sawmills at Gympie and Maryborough, there was little market for cedar in Queensland at the time, and most of it was sent south. The rafts, followed all the way by boatmen, were floated down the coast to Bribie Island, at the northern end of Moreton Bay, and there broken up for shipment to Melbourne by steamer.

It was not only cedar that white settlement quickly cleared out of the stinging-tree country. The forest was full of game and all the creeks had fish in them in those days, but as the cedar-getters, carriers and prospectors moved up they used explosives for their fishing and soon killed every living thing in the creeks; gun clubs were formed in Gympie, and wallabies, turkeys, ducks and pigeons were slaughtered in thousands for sport; large stands of trees were ringbarked so more grass would grow for stock. Before long the whole nature of the country was changed. With no heavy vegetation to protect the ground, the rains, instead of soaking into the soil to keep the springs and creeks running, scoured away topsoil, silted streams and drained away quickly in flash floods to leave dry gullies behind. Land that had once carried enough native game to let hundreds of Aborigines live well became dry forest and cultivated paddocks of the spreading farmlands.

The Mary River, with its large, mountainous watershed barring the path of tropical cyclones, had always been liable to flooding, and every acre of ground that was cleared increased the danger to those occupying it.

Gympie's first serious flood in March 1870 followed several weeks of rain, but as the downpour increased the rising waters of the river still took the young township by surprise. The muddy flood backed up with startling suddenness into the creeks and gullies and spread over the flats where, in the early days of the field, diggers had built to be near the alluvial gold. The Nash's Gully and Chinaman's Flat area disappeared completely beneath the water; so did the whole of the One Mile. Even Deep Creek burst out of its steep banks.

The water poured into hundreds of old shafts, smashed flimsy bark dwellings, and lifted more substantial wooden buildings bodily from their blocks. Foot by foot it crept up Mary Street until it was half-way up Commissioner's Hill. Every building on high ground became a refuge for some of the hundreds driven from the flats. Children slept on church pews pulled together to make bunks. All those hotel lounges, billiard rooms and dance halls that remained above flood level were packed.

The rain stopped at last, and the day the flood reached its peak dawned fine and clear. Soon smoke was rising from hundreds of fires all over the hills as families did the best they could to cook breakfast with wet wood. The low areas were one sea of water through which the course of the fast-flowing river could be detected only by the apparently endless line of cedar logs rushing past on the current.

At one time hundreds of pumpkins came floating past, and hordes of small boys evaded their parents' eyes and risked being whirled away by the current as they swam out into the backwaters to retrieve as many of them as they could.

Farmers who had begun to establish themselves along the river were marooned for days on the tops of their houses with nothing to eat but wet maize and a few pounds of flour.

The Great Flood | 189

Receding waters showed low-lying parts of Gympie swept bare of buildings, mine workings filled in, and debris everywhere. The once thriving separate township of One Mile had disappeared completely. Though later rebuilt to some extent, it was never the same again.

Five years later the town was again flooded, the flight to high ground was repeated, and once again buildings were smashed or carried away down-stream. One that escaped was Mrs Goodwin's hotel which rode out the flood like a ship at anchor. The building was on stumps and the area underneath it was packed with empty casks which lifted it up like buoys as the water rose. Several strong cables firmly fastened to nearby trees prevented it from drifting away.

Other floods followed, including a bad one in 1890, but the worst of them all was the great flood of 1893.

On Saturday, 28 January, torrential rains and gale winds swept the goldfield. They continued all the following week. By Thursday, 2 February, sixteen inches of rain had fallen. Nash's Gully, the lower part of Mary Street, and the One Mile were flooded. The Mary River was eighty feet above its normal level and was rising about ten inches an hour. More than eight inches of rain fell that day, and on Friday it was still raining hard and the river was still rising. Buildings on low ground were submerged and many were carried away. During Friday and Saturday 120 houses were counted floating down the river.

Mines on the Monkland line and at the One Mile were flooded, and the tall smokestack of the Victoria crushing battery rose solitarily from the flood. Everything else was water and mud, desolate except for the deafening cacophony of thousands of frogs.

During Friday a number of mine workings were blown up by the pressure of air trapped in their drives and compressed as the shafts filled with water. Pithead gear was wrecked. At No. 2 and 3 South Smithfield mine the massive bearers and other woodwork, together with pulley wheels and other gear, were carried high into the air and scattered in all directions. So great was the pressure of the imprisoned air in some places that the surface of the ground was heaved up and cracked by it. An eye-witness of some of the explosions described them as being "grand in the extreme, especially that at the No. 1 North Glanmire eastern shaft, a large column of water being raised to a height of over 100 feet and then falling in cascades accompanied by great showers of spray".

The wind slackened on Friday afternoon, but all that night and throughout Saturday and Sunday the flood waters continued to rise, passing levels reached in 1870 and 1890 by more than ten feet. Flooded-out householders and shopkeepers were forced to move higher and higher seeking refuge on the hills that stood as isolated islands in the brown waters.

At 4 p.m. on Sunday the water began slowly to subside, leaving behind it wrecked buildings, machinery and farmlands, huge heaps of debris and vast stretches of stinking mud. The flooding of the mines and the damage done by the explosions put most of them out of operation for months to come, and in doing so, left hundreds of men without work.

A few mines had escaped serious damage and were back in operation again fairly quickly. At the Phoenix P.C. Tribute the water was pumped out and the men returned to work. Some crushing mills were able to resume. But even these were back in operation only temporarily and would soon be forced to close down because of dwindling stocks of coal which came by rail from Burrum in the north and Ipswich in the south. Gympie had been linked by rail with Maryborough in 1881 and with Ipswich in 1891, but now both links were cut, the northern line by the washing away of the rail bridge over the Mary River at Antigua, and the southern by the destruction of the bridge over the Brisbane River at Indooroopilly.

But even as the townspeople began to clean up, the floods were far from over. The night of Friday, 19 February, brought a terrific storm of wind and rain that continued all Saturday. By midday Saturday more than thirty inches of rain had fallen and the river was again pouring its waters over the low ground not yet clear of the previous flood. All day Sunday the water rose, and the weary townspeople were forced to abandon the clean-up and again seek shelter on higher ground. It rained right through until the following Friday. On Saturday, 18 February, the flood began to subside.

At last Gympie was able to take stock of its losses. The *Gympie Times* reported:

> The floods have leaked underground and filled vast underground reservoirs which have been formed by the labours and skill of men during many years.
>
> At least 970 men will be thrown out of work for the next four months, twenty mines are submerged and a vast amount of labour and money must be expended before the mines can resume working operations. The companies closed were keeping 135 head of stampers continually going. They disbursed every fortnight about £4,500, equal to £36,000 before work can be resumed.
>
> Gympie's losses may be summarized: 120 houses wrecked or lost, crops in the neighbourhood totally destroyed, cattle and horses killed, twenty mines flooded, 567 practical miners thrown out of work for from two to four months, 200 workmen usually dependent on the working of the mines also thrown out of work for a similar period, heavy losses to the trading community plus loss of trade, and the death of four persons. Personal losses must be added to this.

The floods that had shattered Gympie were the most widespread and severe in the history of the colony, the trail of devastation stretching from Rockhampton in the north, to far south of the Queensland border and up to a hundred miles inland. Losses had been tremendous.

There followed a financial panic in which some banks closed their doors and others restricted all advances very severely. Few people felt the pinch harder than the people of Gympie, where large sums were needed to put the mines back in operation. Money was not forthcoming.

As in the previous crisis it was the spirit of self-sufficiency in the tightly knit goldfield community that brought it through. Mine owners reached private agreements with their miners. Workmen waited for their wages and accepted what could be paid. As mines came back into production, gold from the early crushings was sent direct to the mint to be turned into sovereigns. Wages and dividends were paid in sovereigns. So great was the resentment against the banks that many men refused to enter a bank and stored their sovereigns in any place that seemed safe, as in the early days of the field.

The private agreements that speeded the goldfield's comeback were made possible because nearly every Gympie miner who had any money to spare was himself a shareholder in at least one goldmine. Many directors of mining companies were themselves working miners. Management and labour were rarely in such a good position to appreciate each other's problems.

But this was not to last much longer. Even as the last mines were being cleared of the mud and rubbish the flood had left there, the times were changing. At the parliamentary polls of 1893 a Gympie miner named Andrew Fisher, a regular speaker at open-air political meetings—and a future Prime Minister of Australia—was elected to the Labor benches of the colony's Parliament as one of Gympie's two members. His fellow member, but political opponent, was William Smyth, pioneer director of the fabulously wealthy No. 1 North Phoenix mine.

Chapter Twenty-one:
Golden Dividends

As Gympie gold gradually came back into production after the flood, new names appeared as big gold producers. They brought the days of the goldfield's peak output—less stirring than the early days of the alluvial, less conducive to surprise fortunes than the days of the first rich reefs, but days of massive quartz crushings and golden dividends far surpassing any prospector's dreams.

Among the new names was that of the Smithfield United, a venture whose development was yet another story of persistence in the face of setbacks and ridicule.

It began in the 1880s with the opening of Gympie's second reefing boom, when Alexander Pollock—the Great Alexander, they called him by then—took up a piece of ground on the Smithfield line of reef so far in advance of all the other claims that miners called it Pollock's Folly. A shaft was sunk, it produced nothing, and the ground was abandoned.

Some years later W. H. Couldrey, who then owned about three-quarters of the shares in the Phoenix P.C., which adjoined Pollock's lease to the west, applied for a twenty-five-acre lease on this line and included Pollock's Folly in its boundaries. In the meantime claims had been taken up on the north of this lease by the North Smithfield Co. and heavy gold yields obtained.

Capital was needed to work Pollock's old shaft, which had been used as a rubbish dump, had filled with water, and come to be called Pollock's Waterhole. But so tight was credit at the time that in spite of Couldrey's extensive shareholdings, he was unable to obtain money from the banks.

Couldrey offered to sell ten acres of the northern part of his lease to the adjoining North Smithfield Co., but without success. Not to be beaten, Couldrey then formed a new company—Smithfield United—and arranged for the Phoenix P.C. to surrender its old lease and take out a fresh one under which Phoenix P.C. gave up ten acres on its western boundary and took in ten acres on its eastern boundary so as to include Pollock's Waterhole. The Phoenix P.C. mine was then let on tribute to another company, the Phoenix crushing battery of forty stampers put to crushing for public hire, and the money so obtained used to buy a second-hand portable steam engine and winding plant which was set up over Pollock's Waterhole. The old shaft was pumped out, cleared of its rubbish, and digging began.

Paying their way from the Phoenix P.C. earnings as they went, Smithfield United went down through the hard rock, drove out their cross cuts, and at last found the reef they were looking for.

The reef paid from the first load of ore, and for the twelve months ended October 1894 the ground that men had called Pollock's Folly produced about a ton of gold worth £101,719. In two and a half years 40,050 tons of ore from the mine yielded 71,139 ounces of gold and paid dividends totalling £169,583.

Couldrey's faith and persistence had earned him another fortune. He later retired to Sydney, bought a large holding in the Port Jackson and Manly Steamship Company, and built a home on Elizabeth Bay overlooking the harbour.

Gympie gold production remained at a high level throughout the 1890s, and in consequence capital flowed in fairly freely for promising ventures. But the day of the battlers was far from over.

The shareholders of No. 2 South Great Eastern had been prospecting the western—and presumably most likely—part of a twenty-five-acre lease on the southern fringe of the field for several years, and paying regular calls for capital without any sign of a profitable return. Many spent all their money on it and had to get out.

Then those who remained took a gamble and started a crosscut from the western shaft out into the unknown ground to the east at a depth of 775 feet. The going was hard and slow, and the stone showed little promise.

As the tunnel extended farther and farther from the shaft, working in it became increasingly difficult. Still the surviving shareholders drove on. The point of no return had been passed; all that was left was gold or ruin. Then they struck a reef. Followed up, it proved rich, but by then the miners were so far from the main shaft that working the reef was almost impossible.

The promise of gold, however, brought capital trickling back. It was used to put down another shaft over the new reef on a site beside the Brisbane Road which was soon to become one of the goldfield's most famous mines.

A number of new reefs were found in drives sent out from this shaft—among them the famous Number Three reef. As Number Three was followed, and the roof of the tunnel cleared, the most amazing display of gold ever seen on the field was revealed. A light shone upwards picked out gleaming white quartz and rich, black plumbago facings held together by bands of gold which even impregnated the rocky casings of the reef. In places the golden network spread right across the top of the level from one wall to the other. The gold seemed to have forced itself into every crack and cranny it could find. It even seamed some pockets of opalescent calcspar—a phenomenon never before encountered.

The No. 2 South Great Eastern began to pay dividends from 1899, but it was not until 1902, with a hundred stampers operating continuously, that yields approached their top figures. In 1902, 33,642 tons of ore yielded 44,571 ounces of gold for dividends totalling £99,000; in 1903, 41,615 tons yielded 52,009 ounces of gold for £115,200; and in 1904, 52,842 tons yielded 49,934 ounces for £113,400. The best single crushing was in December 1903, when 4,920 tons of ore yielded 5,681 ounces of gold for a total dividend of £14,400.

The returns for these years—1902, 1903 and 1904, totalling £327,600—were the highest for any single mine in the history of the Gympie goldfield.

All that remains of the mighty mine today are a few heads of stampers, several cages in which miners descended the shafts, and the old concrete water tank which, remodelled and roofed, now houses the Gympie Historical Museum.

One of Gympie's most famous mines, the Scottish Gympie, owed its existence to a determined Scotsman named Matthew Laird, who worked on the principle that if there was no way of getting round a problem, the only thing to do was hammer right through it.

Matthew's family were timber merchants, shop owners and manufacturers, of Glasgow. After gaining experience in the company he came to Queensland in 1887 at the age of twenty-nine and, after six years and several false starts, found himself married and broke. He had, however, made some useful business connections, including Gympie share-broker Mr H. Willett, and as a result of a plan worked out by the two of them, he left for Scotland in June 1894 with the idea of interesting Scottish investors in the Eastern Monkland, a mine that had until that time produced nothing for shareholders who were anxious to get rid of it. However, a few experienced miners believed it had some promise. The price was only four thousand pounds cash and two thousand fully paid-up shares, but even at that there was little material with which to set about convincing canny Scottish capitalists that it was a good proposition.

In spite of a poor response, Laird kept campaigning, and by March 1895 had formed the Scottish Gympie Gold Mines Co. Ltd. and was on his way back to Queensland with four thousand pounds which the Gympie shareholders were delighted to accept for a mine which they were sure was of no use to anybody. The Scottish company, with Laird as managing director and secretary, took over in April 1896 with the shaft down 670 feet in a locality in which slate could be expected to be reached at about a thousand feet.

The thousand-foot level was reached and passed without any sign of slate. Local men, who had never expected much from the mine anyway, shook their heads. A surveyor recommended driving crosscuts into the surrounding rock to see if anything could be found there. It was a forlorn hope that offered no real chance of success, and Laird rejected it. "So long as we have money left to dig with, we'll go down," he announced, and dig down they did.

The shaft went down to eleven hundred feet, to twelve hundred, to thirteen hundred and to fourteen hundred feet with no sign of slate, and by then none but Laird had any expectation of finding it at all. Then, at fourteen hundred and two feet, they struck it. They followed it, found quartz, and in the quartz, gold.

The Scottish Gympie mine paid its first dividend in December 1898 and production improved steadily until 1906 —its peak year—when 91,700 tons of ore yielded 44,808 ounces of gold for dividends totalling £82,500. Though production figures never reached the peak attained by No. 2 South Great Eastern, its good yields lasted longer, and by 1917—twenty years after its first crushing in 1897—Scottish Gympie had topped the list of Gympie gold producers with a total of more than £2 million worth of gold.

During its peak years, the mine was the largest on the field, employing 330 men, working levels from 2,000 to 2,500 feet deep, with more than thirty miles of underground roadways, and keeping 150 stampers going continuously. Its old retort house—but little else—still stands today as one of the few reminders of Gympie's golden years.

Chapter Twenty-two: Green Pastures

Agricultural development around Gympie was confined at first to tilling the rich river scrublands and cultivating a few good forest plots to provide farm produce and vegetables for the goldfield. Only by the late 1890s was it recognized that the country immediately surrounding the town was good farming land. After that, with the booming goldfield providing a ready market, development was rapid.

By 1903, the peak gold production year, most of the country surrounding the goldfield was being farmed. Gympie, with a population of about thirty thousand, was proclaimed a town. Its streets, still following the lines of the old bullock tracks, had been paved with stone from the mullock heaps, and contained traces of the gold that had brought the town into existence.

Gold production began to fall off during the first decade of the century, but it continued to be the town's main source of income until about 1917—fifty years after Nash reported his discovery. After that diminishing returns and rising costs closed mine after mine.

The goldfield's veteran, No. 4 North Phoenix, which began production in February 1881, closed down in 1925 after forty-four years of operation. It had worked six different reefs and crushed 85,191 tons of ore for a total yield of 128,556 ounces of gold worth £456,373. Its peak production year came close to the end of its run, with a very rich patch of 120 tons of quartz crushed in June 1922 to produce 3,358 ounces, 3 pennyweight, 18 grains of gold.

Its manager, James Brown, was also a veteran of the field. He was one of the original shareholders, became manager in November 1882, and retained the position until the mine closed.

Among the goldfield's record figures were: Wilmot Extended, best crushing of 470 tons of ore for 10,944 ounces of smelted gold. No. 2 South Great Eastern, top dividends of £327,600 for the three years 1902, 1903 and 1904. Scottish Gympie, biggest total production of gold for the field—608,279 ounces from 1,159,162 tons of ore. Gympie's top production year was 1903 with 176,369 ounces of gold.

Gympie's total gold-production figures are uncertain, a number of estimates having been made. But in 1927, by which time production had practically ceased, the field's total output was worked out at 4,084,720 ounces, valued at £14,296,320 (on present day values about $126 million).

By that time farming had done a good deal to replace the lost revenue of the closed goldmines. The Wide Bay Co-operative Dairy Association was formed in July 1906, Golden Nugget butter made its appearance to replace the golden nuggets from the mines, and Gympie's butter factory was on its way to becoming the largest in Australia.

Dairy produce, beef cattle, pigs, fruit and vegetables, and timber became Gympie's staples, and the seven hills from which the red flags once fluttered to announce each new gold find now looked out across green pastures.

Of the booming gold days little remained. All that was left of the mining machinery lay rusting among weeds and tall grass, concrete slabs covered the deep shafts, and some of the copper plates from the stampers had been taken away to make roofs for chicken houses. The last of the mullock heaps were being carted away to build roads and fill in the shovel-scarred creeks and gullies of old Nashville to form growing Gympie's new playing fields and parklands.

But gold can never be quite forgotten, and there were always a few men who fossicked amid the mullock heaps, studied old reports, and formed theories about reefs that might still remain to be discovered.

In September 1969, just over a hundred years after the first rush, Gympie El Dorado Gold Mines Ltd., with a paid-up capital of $150,000, began work amid the gum-trees flanking the city's Albert Park sports oval to put down a shaft to look for a reef that members felt sure lay undiscovered beneath the park. The site was right beside the Bruce Highway leading to North Queensland, and about four hundred yards from the old workings of some of the richest reefs of Gympie's golden days.

Chairman of directors of the Gympie El Dorado was Gympie's former mayor of many years standing, R. N. Witham, who told the Press at the time, "We are putting $97,000 into this hole; about a fifth of the money comes from Gympie, the rest is from America, New Zealand and southern States." The four-man team doing the digging was headed by manager Bill Reeve, a veteran engineer who went to Gympie in 1903 and never stopped looking for gold.

In May 1970 another mine, hopefully named the Phoenix Reborn after a succesful mine of the boom days, was opened in a vacant allotment amid the houses of suburban Ray Street. Miner Stan Smith, head of the six-man syndicate digging it, was confident that there was still gold down there somewhere in the Gympie slates.

At the end of October 1971, with the El Dorado shaft still going down, Mr Witham described prospects as very bright indeed. "A solid reef is in evidence," he said. "It is small, but heavily mineralized, and shows colours of gold."

The stinging-tree gold is not entirely gone.

Bibliography

Among sources consulted in the compiling of this book are the files of the *Gympie Times*, the *Courier* (Brisbane), the *Queenslander*, the *Queensland Government Mining Journal*, and the following books:

Alcazar Press, *Queensland, 1908*. Brisbane, 1900.

Allen, Charles H., *A Visit to Queensland and Her Goldfields*. Chapman and Hall, London, 1870.

Bicknell, Arthur C., *Travel and Adventure in Northern Queensland*. Longmans, Green and Co., London, 1895.

Bird, J. T. S., *The Early History of Rockhampton*. The *Morning Bulletin*, Rockhampton, 1904.

Cilento, Sir Raphael, and Clem Lack, *Wild White Men of Queensland*, W. R. Smith and Paterson Pty. Ltd., 1963.

Coote, William, *The History of the Colony of Queensland*. Brisbane, 1882.

De Satge, Oscar, *Journal of a Queensland Squatter*. Hurst and Blackett Ltd., London, 1901.

Fitzgerald, John D., *Studies in Australian Crime*, First Series. Cornstalk Publishing Co., Sydney, 1924.

Golding, W. R., *The Birth of Central Queensland*. W. R. Smith and Paterson Pty. Ltd., Brisbane, 1966.

Hill, W. R. O., *Forty-five Years' Experiences in North Queensland*. H. Pole and Co., Brisbane, 1907.

Idriess, Ion L., *Prospecting for Gold*. Angus and Robertson, Sydney, 1931.

Kennedy, Edward B., *The Black Police of Queensland*. John Murray, London, 1902.

Lees, William B., *The Goldfields of Queensland*. Outridge Printing Co. Ltd., Brisbane, 1899.

Loyau, George E., *The History of Maryborough and Wide Bay and Burnett Districts*. Pole, Outridge and Co., Brisbane, 1897.

Meston, Archibald, *Geographic History of Queensland*. Edmund Gregory, Government Printer, Brisbane, 1895.

Nimey, Aleck J., *The Gympie Mining Handbook*. Muir and Morcom, Brisbane, 1887.

Bibliography

O'Sullivan, N., *Cameos of Crime*. Jackson and O'Sullivan Pty. Ltd., Brisbane, 1947.

Pattison, J. Grant, *Tales of Early Queensland*. Fraser and Jenkinson Pty. Ltd., Melbourne, 1939.

Petrie, Constance Campbell, *Tom Petrie's Reminiscences of Early Queensland*. Angus and Robertson, Sydney, 1932.

Rands, William H., *Handbook of Excursions, Geology of Gympie and District*. Pole, Outridge and Co., Brisbane, 1895.

Stirling, A. W., *The Never Never Land*. Sampson Low, Marston, Searle and Rivington, London, 1884.

Thackeray, J. R., *Gympie and Its District, A Field for Emigration and Settlement*. Robinson and Co., Maryborough, 1885.

Thorne, Ebenezer, *The Queen of the Colonies*. Sampson Low, Marston, Searle and Rivington, London, 1876.

Wade-Brown, Nugent, *Memoirs of a Queensland Pioneer*. Stencil copy of typescript dated Nov. 1944. In possession of Oxley Memorial Library, Brisbane.

Index

Abbott, H. P., 65
Aborigines, 2, 3, 5, 7, 8, 26, 35, 47, 71, 78, 93-4, 116, 167-70. *See also* Campbell, Johnny
Accidents in mines, 119-21
Agricultural Reserve, 143-5
Agriculture, 168, 170, 188, 198-9
Ahern's premises, 73
Alliance Company, 145
Alluvial gold, total Gympie production, 109
Alma Station, 17
Amamoor Creek, 174
Ambrose, Walter, 88, 170
Antigua bridge, 190
Apis Creek, 20
Aplin, C. D'Oyly H., 115, 162-3
Appolonian Vale Hotel, 74, 90
Archer brothers, 9, 16
Archibald, Alexander, 142-5, 147-9, 153
Argo, George, 182
Atherton, James, 11
Aurora (schooner), 5

Bank of New South Wales, 129
Barlow, Billy, 74
Barnes, Hiram, 133
Barnett, J. J. T., 23, 51
Barrett, Charles, 87
Barron (settler), 5
Bath, William, 155
Bedford (hotel owner), 64
Beer, George, 155
Bella Creek, 32
Bethel Church, 102
Bidwill, John Carne, 6-8, 23
Big Cake, 155-8

Big Toby, 172
Billy (black tracker), 175
Black, Adam, 114
Black, Robert, 114
Blackbird (ship), 117
Black Swamp, 117
Black Swan (ship), 116
Blake, Joseph, 132
Bluey, 150-1
Bond, William, 139
Bonnie Doon Bill, 18-9
Bonnie Doon goldfield, 19
Booker's Currie Hotel, 130
Booth (miner), 105
Booth's Post Office, 73
Bouel, French Charley, 99, 163
Bowen, Governor Sir George, 27
Boyne River, 4
Braddell, Frederick, 155
Bragg (Government Road Overseer), 56-7
Bramston, Hon. J., 161
Breddall's Maryborough Hotel, 100
Bribie Island, 187
Brier, Julian, 164-5
Brigg, Valentine, 83-4
Brisbane, 2, 3, 6, 13, 25, 30, 177
Brisbane Courier, 38
Brisbane Free Press, 13
Brisbane River, 3
Brown, Charlie, 40
Brown, James, 199
Brown, Katie, 20
Brown's forge, 73
Bruce Highway, 200
Buchanan (surveyor), 8
Buckland, C. J., 129
Buckland, John Francis, 133, 135, 138

Bullock wagon travel, 87-8, 91-3, 117
Bunya pine, 8
Burnett, J. C., 4
Burnett River, 4
Burns, J. & J., 73
Burrum River, 39, 149
Burton's Newmarket Hotel, 100
Bushrangers, 128-32, 135-53
Byers, Mr and Mrs J. L., 100
Byrne, Dr Theodore Edgar Dickson, 159
Bytheway's store, 73

Cahill, Trooper, 61-4
Caledonian Hill, 33, 35, 71
Caledonian P. C., 113-15; reef, 112-13, 115
Calliope goldfield, 17, 28, 31, 149
Calloppy, Sergeant, 139
Calton Hill, 71
Campbell, Johnny, 170-4
Campbell, Sergeant, 175
Canny, Michael, 83-4
Canoona, goldfield, 10-6; station, 9, 10
Cartwright, John, 40
Cedar getters, 6, 186-7
Cemetery, 47
Chapple, W. C., 10-11, 17
Charters Towers goldfield, 163
Chinaman's Flat, 106, 188
Chinese, at Crocodile Creek, 21-3; at Gympie, 46-7, 96, 98, 105-8
Christie, Frank. *See* Gardiner, Frank
Church services, 75, 102-3
Cinnabar, 172
Claim jumpers, 81-2
Clarke, Gold Commissioner, 131
Clarke, Tom, 53
Clarke, William, 113, 122
Clermont, 19, 59; gold escort, 60-3
Coach services from Gympie, to Brisbane, 133-4; to Maryborough, 71, 128, 132-3
Cobb and Co., 101, 132-5, 157, 177
Cockburn, James, 55, 94
Cockburn, Mrs James, 91-4
Cockburn, Mrs Tom, 90
Collin, Charles, 83-4
Collis, J. D., 160
Commercial Bank of Sydney, 54, 129
Commissioner's Hill, 71, 188
Convicts, runaway, 3

Coquette (schooner), 11
Cottier, Captain, 30
Couldrey, William, 122-3, 193-4
Craig, Alexander, 19
Cressbrook Station, 3
Critchley, Robert, 160
Croaker's, spirit store, 73; Northumberland Hotel, 101
Crocodile Baker, 95
Crocodile Creek goldfield, 21-3
Crushing machinery, 109, 114, 124
Curra Station, 23, 35
Currey, Constable, 162
Currie Hotel, 130
Curtis, George, 83-5
Curtis Nugget, 84-5

Dart, George W., 7
Davidson, William, 37, 40, 112
Davis, James, 3
Deep Creek, 81, 188
Denman, R. J., 24, 32, 50-1
Diggings at Gympie, 40-2, 74, 78-80, 82, 98
Dodd's reef, 115
Downey, Constable, 170
Drew, Robert, 114
Duckworth, G., 164-5
Duckworth, Reilly, 95
Duke of Edinburgh, 84
Duramboi, 4. *See also* Davis, James
Durundur Station, 3, 6, 7, 9, 29

Eales, John, 3, 4
Eastern Monkland, 196
Eaton, Fred, 117
Edward (schooner), 3
Eel Creek, 141, 150, 152, 174
Eldridge (miner), 131
Ellen Harkins mine, 181
Elliott family, 9
Elliott, Sub Inspector, 22, 65-7, 149
Ellis (publican), 11
Enoggera, 29
Enterprise crushing machine, 117; Quartz Crushing Company, 117
Etheridge goldfield, 163
Everett, William, 6

Farley's Mining Exchange Hotel, 100
Ferry at Maryborough, 50
Fighting, 75-6
Financial crisis, 191-2

Findlay, Tom, 39
Fisher, Andrew, 192
Fire, 121
First Pocket, 114
FitzRoy, Governor, 4
FitzRoy, Lady Mary, 4
Fitzroy River, 4, 9, 10, 13
Flavelle Bros., 28-9, 35
Floods, 82, 186-91
Florence Irving (steamer), 116
Flynn's Hotel, 151
Foos' Mill Hotel, 100
Forman, E., 171
Fraser Island, 2
Freeman, Bill, 114
Freemason's Hotel, 101
Freeston (coach passenger), 135-7
Freights to Gympie, 117
Fulton's Melbourne Hotel, 100
Furber, George, 4, 5

Gardiner, Frank (Frank Christie), 19-20
Gardiner, Miss, 99
Geographic History of Queensland, 51
Gilbert River goldfield, 163
Gilbert's saleyards, 73
Gildea, Trooper, 61-2
Giles, W. C., 23-4
Gillis, Mrs John, 169
Ginginburras, 3-5
Gladstone, 9, 10, 12, 18
Glanmire P.C., 178, 180
Glasgow, James, 155
Gneering (steamer), 133
Gogongo, 64
Gold, discovered: Bonnie Doon 19, Calliope 17, Canoona 10, Charters Towers 163, Clermont area 19, Cowarral 19, Crocodile Creek 21, Etheridge 163, Enoggera 29, Gilbert River 163, Imbil 105, Jimna 105, Kilkivan 105, Mount Morgan 23, Nanango 27, Nash's Gully 33, Palmer River, 163, Peak Downs, 19, Ravenswood 163, Stony Creek 21, Yabba 105; discoverer's reward, 26, 110; prices paid for, 34, 57, 181; Gympie, alluvial production, 109; first traces of, 6-8, 24, 29, 33; production total, 199; reef production, 114-15, 156, 158, 181-2, 185, 194-5, 197-9; report of discovery of, 37

Gold Escort, 57-8, 190, 128, 154-8
Golden Age Hotel, Gympie, 100
Golden Age Hotel, Rockhampton, 144
Golden Bar claim, 108
Goldfields Act of 1874, 160
Goodchap, Frederick, 54, 113
Goodwin, Mrs, 73, 189
Gootchie, 23, 132
Government Savings Bank, 73
Gracemere Station, 9, 10
Graham and Co., 73
Grand Trianon (ship), 16
Greathead's shanty, 47
Great Eastern mine, 182
Great Smithfield mine, 122
Griffin, Johnny, 175-7
Griffin, Thomas John, 59-68
Grigor, William, 91, 133
Gunpowder, 119-21
Gympie, 44, 89, 96-9, 154, 178, 191-2, 198-9; Historical Museum, 196; Stock Exchange, 182
Gympie Creek, 30, 34, 39, 40, 44
Gympie El Dorado Gold Mines Ltd., 199, 200
Gympie-Gympie tree, 1, 6, 7, 92-3
Gympie Mining Handbook, 31, 182
Gympie Times, 31, 191. See also *Nashville Times*

Hall, T. S., 62-5, 144
Halligan, Patrick, 143-53
Hamilton, Doctor Jack, 76-7, 151, 160
Hanley, Detective, 150-2
Hardy, Captain Philip, 10, 14
Harvey, Bill, 114
Hawk (steam tender), 116
Hayes, Jim, 121
Hazlett, Constable, 162
Heilbronn, J., 135
Hendry (carrier), 117
Herbert, Premier R. G. W., 27
Hero (steamer), 116
Hill, W. R. O., 151
Holt, W. H., 140
Horan, Father Matthew, 103-4
Horse and Jockey Hotel, 100
Hotels and shanties, 74-5, 99, 101
Howard, Mrs, 39
Hudson, Tommy, 100
Hulyer (butcher), 73
Humphreys, Frank, 143, 148, 153

Humpy Back Con, 122
Hyman (cheapjack), 73

Imbil, goldfield, 105; homestead, 6; Station, 32
Indooroopilly bridge, 190
Ipswich, 4, 5, 177; railway to Darling Downs, 25

Jackson, Harry, 114
Jackson, Michael, 155
Jack the Frenchman. *See* Brier, Julian
James Nash (cutter), 81
James Paterson (ship), 116
Jane (ship), 16
Jardine, John, 22-3, 60, 148
Jenny Lind (schooner), 9
Jimna goldfield, 105
Joliffe (midshipman), 3
Jonathon (teamster), 92-4
Jones Hill, 122
Jones, Robert, 155
Joseph (assayer), 165
Julian, Sergeant, 60-2, 65
Jumping Doctor. *See* Byrne, Dr Theodore Edgar Dickson

Kabis, 2, 3, 168, 170, 174-5
Kagariu. *See* Campbell, Johnny
Keliher, Jerry, 162
Kennedy, E. B., 57, 107
Keppel Bay, 15-16
Keyser (fisherman), 94
Kidgell, J., 136
Kidner, Frank, 103
Kift, Robert, 54, 113
Kilcoy Station, 3
Kilfeder, Detective, 65-6
Kilkivan, 35, 171; goldfield, 105-6
King, H. E., 56, 69, 81-2, 113, 160
King, Thomas, 135, 138
King, Constable Thomas, 142, 162, 174-7
King, W. S., 135-8
Kipper Creek, 174
Krohmann Cake, 158

La Barte, H., 71, 87, 128, 132
Lady Mary P.C., 53-4, 57, 111, 115; reef, 53, 112-13, 122
Laird, Matthew, 196
Last (station manager), 4

Lawrence, Franklin, 52, 111, 113
Leishman, W., 39, 40, 43-4, 110, 164-6
Leishman, Mrs W., 42, 44, 80, 164-5
Leopolds (troupe), 97
Lion Creek Hotel, 142-3, 145, 147-8
Lloyd, S. J., 142, 152
Long Bill, 151
Lord, Fred and Robert, 122
Low, James, 91, 133
Loyau, George, 4, 5
Luck, Frank, 169

McDonald, A., 50
Mackenzie River, 64
McMahon, Ignatius, 164-5
McMahon, Sergeant, 21
McNamara (miner), 103
Macpherson brothers, 122
Malcolm, Billy, 35-7, 39, 40, 80, 109, 164, 166
Maori mine, 160-1, 165
Markwell's premises, 73
Maroochy River, 133
Marquis of Normanby, 160-1
Martin, Constable, 170
Maryborough, 5, 6, 30, 34, 38, 42, 44-5, 116, 177
Maryborough Chronicle, 43, 53, 58, 100, 105, 116, 119, 128, 130
Maryborough Hotel, 100
Mary River, 2-7, 23-4, 29, 34, 82, 168, 186-91
Mary Street, 57, 70, 72-3, 188-9
Mason, Dr, 131
Mason's pub, 74
May's shanty, 73
Melbourne Hotel, 100
Meston, Archibald, 51
Mining companies, 155
Mining courts, 48, 159-60
Mining Exchange Hotel, 100
Mining legislation, 46, 159
Mining machinery, 118, 180, 182-4
Mooloolah River, 91
Moreton Bay, 3
Moreton Bay Courier, 4
Morgan, Edwin, 54, 113
Morinish goldfield, 144; station, 140
Moss, Joseph, 21
Mother of Gold, 53, 55
Mount Coora, 171
Mount Morgan goldfield, 23

Monkland, P.C., 155; reef, 155, 189
Muir, Gilbert, 114
Muller (miner), 103
Mulligan, James Venture, 77, 151, 160, 163
Murdoch (miner), 81
Murphy, Catherine (Mrs James Nash), 32, 55, 88, 97, 110
Murphy (miner), 122
Myles, Johnny, 132

Nanango, 26; goldfield, 27-8
Nash, James, 2, 28-40, 42, 51, 80, 105, 109-10, 126
Nash, Mrs James. See Murphy, Catherine
Nash, John, 37-40, 80, 109
Nash's Gully, 33-4, 40, 42, 56, 80, 105, 188-9
Nashville, 44, 69-71, 86, 94-6. See also Gympie
Neurum Creek, 174
New Chums, 21-2, 41, 48, 103
Newmarket Hotel, 100
New Monkland reef, 122
New Zealand P.C., 160. See also Maori mine
Nimey, Aleck, 31, 182
Nine Mile Hotel, 50, 132
Noosa, 2, 133, 176
North Caledonian No. 1, 114
North Glanmire, 179; No. 1, 190
North L., 155
North Lady Mary No. 2, 178
North Phoenix Co., 179; No. 1, 180, 182-5, 192; No. 4, 198
North Smithfield Co., 193-4; No. 2, 122
Northumberland Hotel, 101, 132-3, 156
Nuggety Gully, 80
Number Three Reef, 195

O'Connell, Captain Maurice Charles, 9-11, 13, 15, 18
O'Connell Hill, 71
Old diggers and old-timers, 46, 96
Old Jack, 144-50, 153. See also Williams, John
Omeo (ship), 117
One Mile (town), 99, 188-9
One Mile Creek, 69, 78

O'Regan, Bill, 160
Otago mine, 122
Ottley, Mrs, 61

Palatine Hill, 71
Palmer, George, 140-50, 152-3
Palmer, Mrs George, 150-3
Palmer, John, 49, 50
Palmer, Richard, 9
Palmer River goldfield, 163
Parker, R. A., 9
Parkins, Captain, 11
Peak Downs goldfield, 19
Pearen, Jacob, 155, 158-9
Pearen, John, 155
Pearen, Joseph, 155
Perkins, Paddy, 161
Pengelly's toll bridge, 134-5
Perseverance Nugget, 84
Petrie, Andrew, 3
Pettigrew, William, 91
Phoenix Company, 179; P.C., 190, 193-4
Phoenix Reborn, 200
Pie Creek, 141, 174
Pilkington, G. P., 20-1
Pioneer crushing machine, 114-15
Pirate (steamer), 13
Pockley, Tom, 54
Policeman (ship), 30
Pollock, Alexander, 52, 111, 113, 193
Pollock, Robert, 52, 111, 113
Pollock's Folly, 193-4
Port Curtis, 9
Poulton, Edward, 142, 162
Power, Trooper, 61-4
Prince of Wales Hotel, 100
Pye Bachelor and Co., 114

Queensland financial crash, 25
Queensland Daily Guardian, 56
Queenslander, 44-5, 73, 75, 111
Quigly (town crier), 97

Race meeting, 101-2
Railway, to Ipswich, 190; to Maryborough, 190
Ramsay, Hamilton, 13
Ravenswood goldfield, 163
Ray Street, 200
Redcliffe Peninsula, 158
Red flags, 46, 179
Red Hill, 71

208 | Gympie Gold

Reef, gold yields, 114-15, 156, 158, 181-2, 185, 194-5, 197-9; structure, 53, 111-13, 115, 154, 163, 179, 182-4, 195
Reeve, Bill, 200
Rice's premises, 73
Ridgelands goldfield, 145, 149
Ringtail Scrub, 174
Rise and Shine Hotel, 172
Road to Brisbane, 133, 157
Robbers, 126-8. *See also* Bushrangers
Robertson, Dr, 10-11
Rockhampton, 13, 15-16, 18, 20, 22
Rockleigh, 61
Rose (schooner), 43
Rose's restaurant, 73
Ross, W., 172
Royal Hotel, 100
Russell, Henry Stuart, 3, 4
Ryan, Dr, 164
Ryan, Stephen, 155

Sailor's Gully, 40, 80
Salmond, Dr David, 65-6
Saxonia (ship), 116
Scanlan, John, 15
Scott, Bill, 76
Scottish Gympie Gold Mines Co. Ltd., 196; mine, 182, 196-7, 199
Scrubby Gully, 80
Second Pocket, 117
Separation of Queensland from New South Wales, 9, 25
Seven Mile Shanty, 138, 169
Share speculation, 124-5, 182
Sheridan, H. B., 37
Sinclair's Horse and Jockey Hotel, 100
Six Mile Creek, 33
Skyring, Zachariah, 26-7, 88, 141
Slade (member of Bidwill's expedition), 7
Smithfield United, 193-4; reef, 122
Smith, Stan, 200
Smolden, Daniel, 155
Smolden, John, 155
Smyth, Detective, 138
Smyth, Jane, 91
Smyth, William, 192
Southerden, W., 34, 43, 142; store, 73
South Glanmire, 182
South Great Eastern No. 2, 182, 194-5, 199

South Lady Mary, 115; No. 7 and 8, 164
South Monkland No. 7 and 8, 155, 158, 178-9
South Smithfield No. 2 and 3, 189
Stable, J. W. (Wicky), 152-3
Stanley River, 3
Stanton, Mick, 23, 29
Stanwell, 61
Stevens (squatter), 175
Stinging tree. *See* Gympie-Gympie
Stony Creek Goldfield, 21
Struck Oil, 166
Surface Hill, 71
Sykes, Captain A. E., 16

Taylor, Charlie, 143, 147-9, 153
Tewantin, 176
Thompson's Flat, 132
Thrower's Freemason's Hotel, 101
Tiaro, 3, 132, 187
Till, Captain, 30
Tinana Creek, 4-8
Toll gates, 134-5
Tom's store, 73
Toogoolawah, 3
Toomey, Con, 162
Tozer, Horace, 160
Traveston Station, 30, 33
Trodden, William, 132
Trollope, Anthony, 154
True Blue Mine, 122
Two Mile, 47, 114

Uhr, Inspector, 149
Upper Mary River Goldfield, 44

Varieties Theatre, 97, 100
Victoria crushing machine, 124, 189
Victoria House, 158

Wade-Brown, Nugent, 113, 122
Walker, Alligator, 80-1
Walker (miner), 135-6
Walker (town crier), 97
Walker's Gully, 56, 80-1, 108
Walsh, Hon. William, 35, 40
Walsh, Maurice, 40
Wannell, Bob, 32-3
Ware, Sergeant, 37, 40
Water supply, 162
Welcome Nugget, 84
West's public house, 73

Wheatsheaf Hotel, 164
Whim, 118
Whip, 118
White (miner), 114
White's Gully, 45, 56, 80, 114
White, R. H. D., 129-30
Wide Bay, 4
Wide Bay Cooperative Dairy Association, 199
Wide Bay River, 4
Widgee Station, 30; crossing, 87
Willett, H., 196
Williams, David, 18-9
Williams, John, 143-5, 153
Wilmhurst, William, 5

Wilmot Extended mine, 181, 199
Wilson and Co., 73
Wiseman, Police Magistrate, 153
Witham, R. N., 200
Women, first on diggings, 43-4; objects of curiosity, 89, 90
Woodrow and Scott, 121
Woods, Constable Frederick, 11-12
Woombye, 157
Wrottesley, W., 3

Yaamba, 15
Yabba, Creek, 6; goldfield, 105; station, 28, 31
Yarra Yarra (steamer), 13, 30

ALSO BY HECTOR HOLTHOUSE

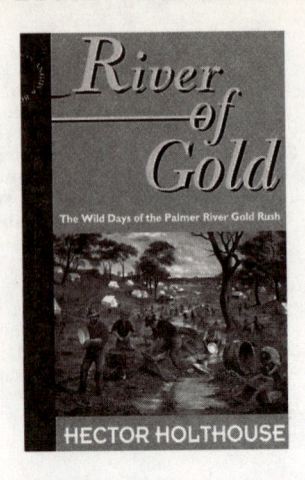

RIVER OF GOLD

The first of his Australian histories, *River of Gold*, is an account of the Palmer River gold rush — sparked off in 1873 by a find of 102 ounces of gold in a few weeks, and over 100 tons in a few sensational years.

"The gold rush at Palmer River, on Cape York, lasted about seven years in the 1870s, but with 35,000 diggers it was this country's wildest while it lasted. Holthouse has researched the story of those days well to make a lively and very readable book."
The *Bulletin*

ISBN 0-207-18778-9

ALSO BY HECTOR HOLTHOUSE

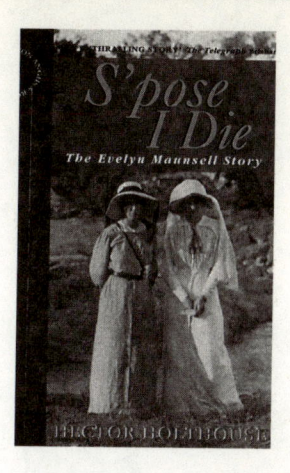

S'POSE I DIE

"This English girl will never stick it out," said one of the bridegroom's friends when Evelyn Evans arrived in Cairns in 1912 to marry Charles Maunsell. She went from a comfortable house near London to an isolated Mount Mulgrave homestead with an unlined roof and antbed floors. For months in the wet season the station was cut off from the outside world, and more than once in the lonely weeks when the men were away mustering Evelyn Maunsell came near to death from illness or marauding Aborigines.

S'pose I Die is set in the same country as depicted in *River of Gold*, after the rush was over and beef had replaced gold as the main export. It is based on Eve Maunsell's written recollections and on her conversations with Hector Holthouse about her life in the Mitchell River country and on the Atherton Tableland.

"An enthralling story"
The *Telegraph*, Brisbane
"A lively and readable account of a pioneer woman's life"
The *West Australian*, Perth

ISBN 0-207-18764-9